HIGH-SCHOOL BIOLOGY TODAY AND TOMORROW

Papers Presented at a Conference

WALTER G. ROSEN, EDITOR

Committee on High-School Biology Education
Board on Biology
Commission on Life Sciences
National Research Council

NATIONAL ACADEMY PRESS
Washington, D.C. 1989

National Academy Press • 2101 Constitution Avenue, N.W. • Washington, D.C. 20418

NOTICE: The project that is the subject of this report was approved by the Governing Board of the National Research Council, whose members are drawn from the councils of the National Academy of Sciences, the National Academy of Engineering, and the Institute of Medicine. The members of the committee responsible for the report were chosen for their special competence and with regard for appropriate balance.

This report has been reviewed by a group other than the authors according to procedures approved by a Report Review Committee consisting of members of the National Academy of Sciences, the National Academy of Engineering, and the Institute of Medicine. The National Academy of Sciences is a private, nonprofit, self-perpetuating society of distinguished scholars engaged in scientific and engineering research, dedicated to the furtherance of science and technology and to their use for the general welfare. Upon the authority of the charter granted to it by the Congress in 1863, the Academy has a mandate that requires it to advise the federal government on scientific and technical matters. Dr. Frank Press is president of the National Academy of Sciences.

The conference High-School Biology: Today and Tomorrow was organized by the Board on Biology of the National Research Council's Commission on Life Sciences. The views in this book are solely those of the individual authors and are not necessarily the views of its sponsors.

Support for the publication of these papers was provided by the Howard Hughes Medical Institute, Bethesda, Maryland.

Library of Congress Cataloging-in-Publication Data

High-school biology : today and tomorrow / Committee on High-School Biology Education, Board on Biology, Commission on Life Sciences, National Research Council.

p. cm.

Papers from a conference held Oct. 1988, in Washington, D.C.

ISBN 0-309-04028-0

1. Biology—Study and teaching (Secondary)—Congresses.

I. National Research Council (U.S.). Committee on High-School Biology Education.

QH315.H615 89-13141

574'.071'2—dc20 CIP

Cover photos: (top) Photograph by James Sherwood; (bottom) Microphotograph of the alga, *Volvox aureus*. Courtesy, National Science Teachers Association.

Printed in the United States of America

COMMITTEE ON HIGH-SCHOOL BIOLOGY EDUCATION

TIMOTHY H. GOLDSMITH (*Chairman*), Yale University, New Haven, Connecticut
CLIFTON POODRY (*Vice Chairman*), University of California, Santa Cruz, California
R. STEPHEN BERRY,* University of Chicago, Chicago, Illinois
RALPH E. CHRISTOFFERSEN, Smith Kline and French Laboratories, King of Prussia, Pennsylvania
JANE BUTLER KAHLE,* Miami University, Oxford, Ohio
MARC KIRSCHNER, University of California, San Francisco, California
JOHN A. MOORE, University of California, Riverside, California
DONNA OLIVER,* Elon College, Elon College, North Carolina
JONATHAN PIEL, *Scientific American*, New York, New York
JAMES T. ROBINSON,* Boulder, Colorado
MARY BUDD ROWE, University of Florida, Gainesville, Florida
JANE SISK, Calloway County High School, Murray, Kentucky
DAVID T. SUZUKI, University of British Columbia, Vancouver, British Columbia, Canada
WILMA TONEY, Manchester High School, Manchester, Connecticut
DANIEL B. WALKER, San Jose State University, San Jose, California

Special Advisors

JOHN HARTE, University of California, Berkeley, California
PAUL DEHART HURD, Palo Alto, California

Former Members

EVELYN E. HANDLER (*Chairman*, 1987-1988), Brandeis University, Waltham, Massachusetts
MICHAEL H. ROBINSON (1987-1988), National Zoological Park, Washington, D.C.

National Research Council Staff

JOHN E. BURRIS, *Study Director*
DONNA M. GERARDI, *Staff Associate*
WALTER G. ROSEN, *Consultant*
NORMAN GROSSBLATT, *Editor*
LINDA D. JONES, *Senior Secretary*

---

*Member, Conference Program Committee

BOARD ON BIOLOGY

COMMISSION ON LIFE SCIENCES

The National Academy of Engineering was established in 1964, under the charter of the National Academy of Sciences, as a parallel organization of outstanding engineers. It is autonomous in its administration and in the selection of its members, sharing with the National Academy of Sciences the responsibility for advising the federal government. The National Academy of Engineering also sponsors engineering programs aimed at meeting national needs, encourages education and research, and recognizes the superior achievements of engineers. Dr. Robert M. White is president of the National Academy of Engineering.

The Institute of Medicine was established in 1970 by the National Academy of Sciences to secure the services of eminent members of appropriate professions in the examination of policy matters pertaining to the health of the public. The Institute acts under the responsibility given to the National Academy of Sciences by its congressional charter to be an adviser to the federal government and, upon its own initiative, to identify issues of medical care, research, and education. Dr. Samuel O. Thier is president of the Institute of Medicine.

The National Research Council was organized by the National Academy of Sciences in 1916 to associate the broad community of science and technology with the Academy's purposes of furthering knowledge and advising the federal government. Functioning in accordance with general policies determined by the Academy, the Council has become the principal operating agency of both the National Academy of Sciences and the National Academy of Engineering in providing services to the government, the public, and scientific and engineering communities. The Council is administered jointly by both Academies and the Institute of Medicine. Dr. Frank Press and Dr. Robert M. White are chairman and vice chairman, respectively, of the National Research Council.

Preface

In the spring of 1988, the Board on Biology and its parent body, the Commission on Life Sciences of the National Research Council, initiated a study of the state of high-school biology education. The recognition that things are amiss had been developing for some time, and the legal disputes over the teaching of evolution several years ago had sharpened the Board's sense of the complexity of the problem. The timely and generous financial support of the Howard Hughes Medical Institute allowed this study to begin.

The Committee on High-School Biology Education, consisting of scientists and educators, is undertaking the study, and its report will be issued soon. One of the committee's first tasks was to organize a conference to provide extensive background information that would inform its deliberations. A program committee identified general subjects to be addressed: objectives of biology education and measurement of achievement, curriculum perspectives and content, instructional procedures and materials, teacher preparation, institutional barriers, and implementation. Each was examined by a panel of speakers. Each panel was chaired by a committee member, who provided brief opening comments. The goal of the conference, held in October 1988, was so effectively realized that we feel that the papers given should be available to a wider audience; they appear in this volume as given at the conference, lightly edited for consistency of presentation. The papers embody the research and opinions of their authors and do not reflect the opinions or judgments of the Committee on High-School Biology Education or the National Research Council.

Invitations to attend the conference were extended to teachers and administrators from across the country, and many teachers traveled considerable distance at their own expense to attend. On behalf of the committee, I would like to acknowledge the devotion and professionalism exhibited by that sacrifice. Whatever the failings of our educational system (and, as this volume attests, the failings are many), there is a cadre of dedicated teachers who remain our best hope for change.

In organizing the conference, we indicated our wish to hear not only from panelists, but also from the audience. In addition, the audience was invited to submit written comments to the committee for consideration after the conference. By the middle of the first day, it had become clear that insufficient time had been allotted for audience participation, and the ensuing spontaneous and heartfelt demonstration of frustration from many of the teachers in the audience stimulated us to adjust the schedule. Unfortunately, this volume cannot reflect the long and fruitful evening shared by teachers and members of the committee, starting with dinner and not ending for many of us until the following morning. That informal session punctuated dramatically—and in a manner that cannot be conveyed by chapters in a book—not only the smothering conditions under which many teachers work, but the dedication and imagination that the very best teachers still manage to bring to their profession.

Walter Rosen recruited the speakers according to the objectives outlined by the program committee. Donna Gerardi, Barbara Christensen, and Linda Jones provided essential planning and logistical support, and Norman Grossblatt and Walter Rosen prepared the papers for publication.

Since the conference in October 1988, Evelyn Handler, the committee's original chair, has found it necessary to resign from the committee, and I have replaced her. We all thank her for her early stewardship of the study, which was so well launched with the conference.

Timothy H. Goldsmith, *Chairman*
Committee on High-School Biology Education

Contents

PART I
OPENING ADDRESS AND RESPONSES

PART II
OBJECTIVES OF BIOLOGY EDUCATION AND MEASUREMENT OF ACHIEVEMENT

PART III
CURRICULUM: PERSPECTIVES AND CONTENT

PART IV
INSTRUCTIONAL PROCEDURES AND MATERIALS

PART V
TEACHER PREPARATION

PART VI
ACCOMPLISHING CURRICULAR CHANGES—INSTITUTIONAL BARRIERS

Part VII
Accomplishing Curricular Changes—Implementation

HIGH-SCHOOL BIOLOGY TODAY AND TOMORROW

PART I

Opening Address and Responses

1

Opening Address

EVELYN E. HANDLER

At its first meeting, in April 1988, the National Research Council (NRC) Committee on High-School Biology Education put forth a seemingly straight-forward question: How do we modernize curriculum to keep up with the explosion of knowledge in the field of biology? Not surprisingly, behind that simple question lies great complexity. The distinguished academic biologists on our committee and the scientist-advisers to our sponsor, the Howard Hughes Medical Institute, as well as the teachers and administrators who serve on our committee, recognize *what* a tangled subset of issues the question unleashes. We cannot solve the problems of content without addressing the entire context—or what I choose to call the ecology—of education.

Some of the subset issues that need to be addressed include teacher preparation, instructional objectives and strategies, texts and other instructional materials, institutional context, social context, and developmental factors. And we need to consider the interconnectedness of biology with the other sciences—physics and chemistry, but also earth science and the social sciences. If we consider biology a component of scientific literacy, which in turn is an ingredient of cultural literacy, how do we make our young people literate?

Evelyn E. Handler is the president of Brandeis University. She holds a Ph.D. in biology from New York University and is a former dean of the Division of Sciences and Mathematics, Hunter College, Columbia University. She is also a former president of the University of New Hampshire.

We know we are failing to do so. I could recite a litany of reports and studies that document the dimensions of our failure, but you know them as well as I do. So let me quote from the succinct summary of Armstrong and co-workers (Education Commission of the States, 1988):

> Assessments have shown that the achievement of American students in science has, in general, declined since 1972 and remains poor in comparison to student achievement in other developed countries. Research conducted in the 1970s and 1980s has demonstrated that science instruction has had low priority. It has been, at best, textbook-driven and focused on content. Too often, teachers of science are inadequately trained, and there are shortages of teachers in fields such as physics and chemistry. Enrollment in high school science courses has fallen. Moreover, science textbooks have been heavily criticized as covering too many topics far too superficially. There is, as yet, no consensus on why science should be taught, what should be taught, who should study science and how science education can be changed.

Our youngsters are deficient in their understanding of biology, both as a coherent discipline and as a body of knowledge. Most of them, throughout their lives, will have little ability to relate what they may learn about biology to the world in which they live. But this is not a failure of our children. It is a failure of public policy to acknowledge the living realities of biology . . . the dynamic processes of nature that course through us and around us as creatures of the planet Earth.

If we are going to incorporate biology into the mainstream of cultural literacy, we must think about how biology and technology interact to affect our lives and even our survival as a species. This presents some fundamental problems. How do we deal with the implications of an exploding body of scientific knowledge, such as genetic engineering and the chemistry of the brain? How can we communicate the implications of rapid developments to large numbers of youngsters? Since the time available for instruction cannot expand to accommodate the growth of knowledge, adjustments must be made. What to drop and what to keep? Should we try to be all-inclusive and contend with textbooks of 1,000 pages weighing 20 pounds, and leave it to teachers and administrators to set priorities? And if so, will the teaching of biology then be rational and relevant? Is it now rational and relevant?

There is the problem of coping with our changing planet—global warming, drought, famine, pollution of the earth and seas. We know the epidemiology and complications of the spread of the AIDS virus. How do we incorporate these into our learning objectives, our evaluation procedures, our teacher training, and our texts? More important, should these matters be made a part of the curriculum content, or should we retain the traditional disciplinary perspective of biology?

These are questions of content that are bound up with context. I believe that in order to determine content, we must first articulate the objectives of a high-school biology education. Only when we know our objectives can we develop a strategy for implementing a curriculum.

We shall begin our panel deliberations, then, by addressing the topic of objectives and how they are to be reflected in our evaluation procedures. Let me start by posing some larger questions, in the hope of stimulating and focusing our thinking.

So let us begin!

What do we want to impart to all students about factual information, perspectives on the living world, reasoning skills, and science as a *process*?

How effectively can we measure the attainment of these objectives? Do standardized tests dictate curriculum content? Are there alternative and more sensitive measurements of achievement?

To what extent do texts and other instructional materials drive the curriculum? How does the teacher's own education shape his or her teaching style and objectives?

A question that has always interested us as teachers: what is the effect of the student's prior education on what he or she learns in the biology course? How much biology is taught in other courses, such as health education or earth science, and how much is learned or mislearned from television?

How much biology should be a part of general science? If biology is presented as a discipline, where and how will the student learn the physics and chemistry that underlie biological phenomena?

To what extent should biology focus on social impacts and technological applications? In a world experiencing snowballing environmental crises, what priority should be given to the concept of the biosphere as a life-support system for human survival?

Should the teaching of biology be insulated from religious, political, or social trends and values?

Of what value, if any, are out-of-classroom instruction and experiences? Museums, zoos, botanical gardens, television documentaries, and other formats present innovative opportunities for instruction. Do we use these resources effectively? When we plan and evaluate the classroom experience, should we factor in children's exposure to informal education? Or, since science illiteracy is rampant, should we conclude that informal education is ineffective and therefore irrelevant, and ignore it?

What does cognitive psychology have to tell us about defining our objectives, and about strategies to achieve our objectives? By ignoring the limitations of cognitive development on learning capacity, do we doom ourselves to frustration, if not defeat?

Shayer and Adey (1981) in England concluded from their extensive tests and studies that "there is a massive mismatch in secondary schools between the expectations institutionalized in courses, textbooks and examinations and the ability of children to assimilate the experiences they are given." This issue will be addressed in one or more of our panels. How entrenched is this problem in our classrooms and how can we go under, around, or through learning obstacles?

And last, should the first biology course serve as a recruiting ground for future scientists? Are we adequately serving the needs of students who show a natural affinity for science? Are we ensuring that a new stream of recruits move into teaching and research careers? What can special science schools tell us about educating the talented student?

While our inquiry is wide-ranging, it cannot address all the contextual problems in any detail. We have not scheduled sessions to deal with the special problems of minority-group students from underprivileged backgrounds or the differences in the educational needs of college-bound and non-college-bound students. We also are not explicitly addressing the allocation of time between biology and the other sciences or among subtopics within biology, such as ecology; metabolism; cell, tissue, and organ systems; and plants, animals, or systematics. However, these problems are of concern to the committee, and we hope to hear more about them in the broader context in which biology is taught.

I would like to draw a brief picture of the historical background against which we are undertaking our task. The biology curriculum, as we know it, first emerged at the end of the last century. To this day, most texts and curricula reflect the survey-of-the-discipline pattern established by T. H. Huxley in 1890 in what is generally viewed as the first general biology text (Huxley and Marten, 1892). From the earliest years, concerned groups and individuals have analyzed and criticized biology education. They have struggled to define its objectives and identify appropriate instructional strategies and materials. In a thoughtful article, "Biology Education in the United States During the Twentieth Century," Mayer (1986) reviewed the many major studies. Drawing on Paul DeHart Hurd's (1961) study, *Biological Education in American Public Schools, 1890-1960*, Mayer tells us that most of what we strive for in biology education has been sought for a very long time. A 1909 report from the High School Teachers Association of New York supported an emphasis on applied biology and training in living and recommended such topics as conservation, health and nutrition, ecology, and critical thinking about biology as applied to daily life.

In 1914, a committee of the Central Association of Science and Mathematics Teachers set out as *the purposes of science education* "a knowledge of the world of nature in relation to everyday life, and an emphasis on career preparation and choice, on problem solving, and on a consideration of the degree of credibility of scientific knowledge." And in 1915, a committee on natural sciences of the National Education Association stated such objectives as development of the powers of reasoning and observation and acquaintance with the environment, with the structure and function of the human body, and with biological principles arising from these studies.

The National Academy of Sciences and the National Research Council are no strangers to the century-long effort to improve high-school biology education. By far the most ambitious and influential effort at improving high-school biology education was, and is, the Biological Sciences Curriculum Study (BSCS). Its history, objectives, personae and products are well known to us. There are enough BSCS veterans and current activists in the audience and on our program to ensure that the BSCS's contributions will not be neglected in our sessions. In fact, before our committee members write their report and make recommendations for curriculum content, they might do well to review the themes that pervaded all BSCS textbooks (yellow, green, blue, and those unwritten) and to determine whether any of these need to be amended, replaced, or augmented:

- Change of living things through time: evolution.
- Diversity of type and unity of pattern among living things.
- The genetic continuity of life.
- Growth and development in the individual's life.
- The complementarity of structure and function.
- Regulation and homeostasis: the preservation of life in the face of change.
- The complementarity of organisms and environment.
- The biological basis of behavior.
- The nature of scientific inquiry.
- The intellectual history of biological concepts.

And one more, added by current BSCS Director Joseph McInerney (1987):

- Relationship between science and society.

Before we address these themes, we must ask why the impact of BSCS diminishes and student performance continues to decline in the face of excellent instructional material prepared and field-tested by teachers and scientists who were guided by widely endorsed objectives. Mayer (1986) points out some of the problems: Despite the resounding triumph of the BSCS effort—adoption by over half the nation's school districts, improved student performance, textbook sales in the millions, adaptations by 14 foreign countries—the sad truth is that there is resistance and resentment by the publishing community, by much of the professional academic education community, by many teachers who were unprepared to meet the demands of these new curricula, and by other institutional entities to this brave new approach. Guided by Mayer's analysis of the impediments to implementation of BSCS biology, we will spend a substantial portion of time on strategies for removing institutional barriers.

Implementation, however, becomes a problem only when we have something to implement. So let us think creatively about our task of redefining or restating high-school biology objectives.

Knowledge about the living world and how it works is growing at an increasing rate while humankind's scientific literacy is falling behind. At the same time, our biotic kingdom is deteriorating. The last summer was calamitous. All along our northeast coast, medical waste and coliform bacteria contaminated the beaches. Algal blooms alter marine life. Toxic gases choke our cities. Drought and heat destroyed millions of acres of forests and crops. Was this a statistical blip or part of a pattern of global warming resulting from ozone depletion? We ask ourselves, is nature striking back? Have we exceeded our planet's ability to absorb our abuse? Is the booming global population, with its exponential consumption of energy and production of waste, threatening life as we know it?

If life as we know it is threatened, we must examine every aspect of our human behavior for its impact on nature. Nature *must* be protected, not only for its own sake, but so that in turn it can continue to support human life.

Should the biology curriculum not be seen in that context? Should we not be teaching the biology of survival on the basis of ecology, including human ecology?

In *The Thanatos Syndrome*, novelist Walker Percy (1987) has his hero observe that "this is not the age of enlightenment but the age of not knowing what to do." Not knowing what to do is no excuse for concluding that we can do nothing. We cannot sit by helplessly while biology education continues to fall short of the demands we can and must put on it to address our planet's integrity. We must not give in to despair, but must keep trying to find out what to do. Harold Morowitz, member of the NRC's Board on Biology, which is overseeing our study, is fond of saying, "Optimism is a moral imperative." So let us now, with optimism, get on with the task of figuring out what to do.

REFERENCES

Education Commission of the States. 1988. The Impact of State Policies on Improving Science Curriculum. Denver, Colo.

Hurd, P. D. 1961. Biological Education in American Public Schools, 1890-1960. Washington, D.C.: American Association of Biological Sciences.

Huxley, T. H., and H. N. Marten. 1892. (Rev.) Practical Biology. London: Macmillan.

Mayer, W. 1986. Biology education in the United States during the twentieth century. Quart. Rev. Biol. 61:481-507.
McInerney, J. D. 1987. Curriculum Development at the Biological Sciences Curriculum Study. Educ. Leader. 44(4):24-28. December 1986/January 1987.
Percy, W. 1987. The Thanatos Syndrome. New York: Farrar, Straus & Giroux.
Shayer, M., and P. Adey. 1981. Towards a Science of Science Teaching. Curriculum development and curriculum demand. London: Heinemann Educational Books.

2

Changing Conceptions of the Learner: Implications for Biology Teaching

AUDREY B. CHAMPAGNE

A quarter-century has elapsed since the scientific community last turned its attention to school science. The overriding concern of academic scientists is that once again the content of school science is out of date. Indeed, major developments have occurred in the sciences that are not yet reflected in science textbooks. However, simply updating the content will not adequately raise the quality of school science or significantly improve America's scientific literacy. Attaining these goals requires attention to the nature of instruction, as well as the content of the school science curriculum. As we turn our thoughts to the future of high-school biology, we must not lose sight of the fact that in the last 25 years other significant changes have occurred that should determine in large measure how the new science is taught and whether it is learned. Among these changes are several that should guide our thinking about the nature of science instruction.

Of the many factors that should influence instruction, none is so

Audrey B. Champagne, senior program director in the office of Science and Technology Education at the American Association for the Advancement of Science (AAAS), directs the National Forum for School Science and the Project on Liberal Education and the Sciences. Dr. Champagne was a senior scientist and project director at the Learning Research and Development Center and research professor of education at the University of Pittsburgh before joining AAAS in July 1984. She holds a B.S. and M.S. in chemistry from the State University of New York, Albany, an Ed.M. in science education from Harvard University, and a Ph.D. in education from the University of Pittsburgh.

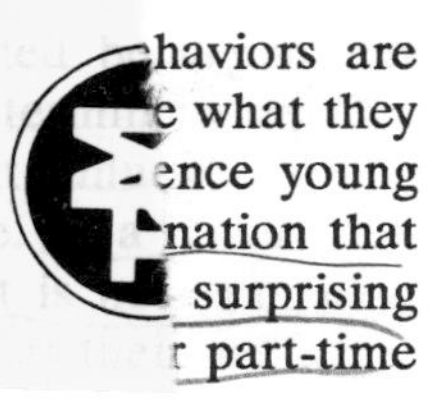

important as the learner. Young people's school-related behaviors are determined by social and psychological factors, which determine what they will learn. Society's values are one of the factors that influence young people's attitudes toward education and learning science. In a nation that values cars, clothes, and cocaine more than learning, it is not surprising that many of our high-school students spend more time on their part-time jobs than on their homework.

Beyond the influence of social values on students' attitudes toward education and learning in general, social values exert profound influence on science learning. The overt manifestations of society's values are public attitudes toward science that are a study in contradictions. At a time when states are mandating more science credits for high-school graduation, society is delivering a contradictory message to American youth regarding the value of studying science. While Americans value the many ways in which science has improved their lives, they are becoming increasingly concerned by environmental degradation and troubled by the difficult moral and ethical choices science places on them. These concerns contribute to negative public attitudes toward science. These negative attitudes are reinforced by the ways in which scientists are portrayed in the media. Many young people have never had personal contact with a scientist. They get their image from the media, which portray scientists as nerds in white laboratory coats with thick glasses who relentlessly pursue science, neglecting family and personal needs. This unappealing image turns young people from science.

Society's image of the scientist presents an even more serious problem for young women, Hispanics, and blacks. Society's perception that science and technology professions are the purview of the white male leads these young people to conclude that science is either socially unacceptable or intellectually unattainable to them. This perception pervades schools and science classrooms, where circumstances in this regard have not changed significantly since I was in junior high school and the science club was for boys only. Today, the message is delivered in more subtle ways—for example, girls don't get called on or answer questions as much as boys in science—but the message is effective.

These comments only touch the surface of the impact of social factors on students' opportunity to learn science and on their choices to study it. There is evidence that for young people from some subpopulations, black and Hispanic in particular, there is a mismatch between the modes of thought of their culture and those of science. In addition, the modes of teaching and learning that these youth experience in the home differ from the modes that they experience in their schools (Cohen, 1986). Such factors as these are social in origin, but have implications for science learning. Science teachers expect that all entering students have the same thinking and learning skills. Such work as that of Cohen illustrates that this assumption is incorrect and places these young people at risk in science. There is also evidence that thinking modes of white females and economically disadvantaged white males have characteristics in common with those of boys and girls in the black and Hispanic subcultures.

The impact of these social factors is particularly important, since the ethnic composition of society is changing rapidly. By the time new curricula are in the schools, the majority of students in classrooms will be children from populations whose members traditionally have not succeeded in science. Achieving a nation populated with citizens who understand and value science requires that these young people learn science.

Other psychological factors also influence science learning, but are not limited to specific subcultures. One of these factors relates to a basic human drive to understand the natural environment. Over the last decade, an impressive body of empirical evidence has developed that demonstrates the consequences of this drive. When children enter kindergarten, they have developed their own private explanations for the events in the natural world. In many cases, these explanations are quite different from those taught in science classes. There is compelling evidence that these ideas are not easily changed by traditional methods of teaching science. Studies of college students, conducted in the United States and overseas, show that ideas developed from experiences with the natural world persist even in students who study science at the postsecondary level and earn good grades.

For example, studies conducted at the University of Pittsburgh and at the Johns Hopkins University demonstrate that exposure to Newtonian mechanics does not result in physics students' giving up their Aristotelian perspectives on the motion of objects. The ideas that heavier objects fall faster than lighter ones and that an object traveling in a circular path will continue in a circular path even after the force moving it is removed are retained even by successful students (Champagne et al., 1980; McCloskey et al., 1980).

Studies conducted in the United Kingdom and Australia illustrate that, just as intuitive Aristotelian thought is resistant to traditional physics instruction, so too is Lamarckian thought resistant to traditional biology instruction. First-year medical students in the United Kingdom and Australia "extrapolate from changes seen within the life time of an individual to account for changes seen in populations selected over many generations" (Brumby, 1984).

Data for these studies were collected during one-on-one interviews. The struggle that students have in reconciling what they believe to be true with what they have been told in science class is evident from the verbal protocols. One task used in the physics studies involves showing the student two cubes of the same dimensions—one aluminum, the other plastic. The student is told that the cubes will be dropped about a meter and is asked to make a prediction about how the times for the cubes to reach the floor will compare. About 75% of the students predict that the aluminum cube will fall "a lot faster." After watching the cubes fall, most of the students who predicted that the aluminum cube would fall faster observed that it, indeed, fell faster—but only a *little bit* faster (Champagne et al., 1980).

One task used in the natural-selection studies poses questions about a photograph of a fair-skinned child of Scottish origins (Brumby, 1984).

Question: If this little [fair-skinned] girl grew up in Africa, what would you predict would happen to the color of her skin?
Student: She'd get sunburnt, then tanned.
Question: If she then married someone of her own race and they lived in Africa and had children born in Africa, what would you predict their children's skin would look like at birth?
Student: (Pause.) The kids could be *slightly* darker at birth.

The extent of the intellectual struggle is only hinted at in the language of the responses. The aluminum block falls only "a *little bit* faster." "The kids could be *slightly* darker at birth." The strength of the personal convictions is particularly well illustrated in the physics example. The students have hefted the blocks. They have observed them fall. And in spite of the fact that under the conditions of the experiment there is no observable difference in the falling time, the students persist in the observation that the aluminum cube is a *little bit* faster.

We should not blame our students for not readily giving up their intuitive notions. While their conceptions of falling bodies and natural selection are not the well-structured formal theories of Aristotle or Lamarck, the history of science illustrates how compelling these ideas are and the difficulty in bringing about change in the scientific community's perceptions about them. This characteristic led Niels Bohr to observe that the scientific community's ideas change only when old scientists die and are replaced by younger scientists with new ideas.

Personal theories are not the only factor that makes science learning so difficult. A project conducted by Sheila Tobias (1986) at the University of Chicago demonstrated that seasoned scholars in disciplines other than physics had problems understanding physics lectures presented by highly skilled physics professors. Among the difficulties they reported was their inability to infer the intent of the demonstrations that accompanied the lectures. They were unsure which observations were important to the logic of the argument that the lecturer was developing. The scientific knowledge base of these highly intelligent individuals is minimal, and that makes the interpretation of the physics lecture difficult. Clearly, the lecturers were overestimating the extent of their colleagues' scientific knowledge and underestimating its importance to their colleagues' ability to understand the lecture.

Social and psychological factors have profound influence on how students interpret demonstrations, lectures, and science text and the extent to which students learn from these experiences. How, then, can biology be taught in a way that will bring about the desired conceptual change, as well as attending to the other purposes of teaching biology in the high school and college?

A necessary condition is that all science teachers at all levels, including college faculty, recognize that students bring personal theories about the natural world to the science classroom. Faculty at more advanced levels cannot continue to blame students' poor understanding solely on the quality of instruction received at the feet of earlier teachers. We have all heard our colleagues at the college level express the wish that the high schools just teach the kids mathematics and leave their minds as like virgin fields prepared to receive their gems of wisdom. By the same token, high-school teachers lament the strange ideas that students develop in junior-high science and, like their university colleagues, wish for students whose minds are empty of all biological theory.

Given that these wishes cannot be fulfilled, how should science teachers proceed? Science teachers must attend to ideas about the natural world that students bring to science class. A consequence of ignoring them is that personal theories remain unexamined and uncoordinated with the canonical theories that teachers present. Traditional instruction is based on the assumption that students' minds are empty vessels to be filled with the knowledge products of the discipline. Canonical ideas are transmitted to the presumed empty vessels by the spoken and written word, with no opportunity for critical examination by the students. Rather, students' minds are more like vessels partially filled with oil. Under the conditions of traditional instruction, water added to the vessel does not mix with the oil. A permanent mixture of the oil and water requires agitation of the vessel in the presence of an emulsifying agent. Agitation is achieved in the classroom via the scholarly interaction of students. The emulsifying agent is supplied by the teacher, who sees that the interactions occur according the tenets of the scientific community.

This strategy is not new. In essence, it is the technique employed in graduate training in the sciences or undergraduate education at prestigious undergraduate institutions. This technique brings together small groups of students under the guidance of a mentor to examine a problem or an idea. Individuals in the group engage in discussion. Initially, each presents his or her own perspectives, which presumably do not match with those of others in the group or, in the case of scientific theory, with canonical interpretation of the theory. Disagreements are argued as participants challenge each other's lines of argument, assumptions, and evidence. The mentor models modes of argumentation, challenge, and application of rules of evidence characteristic of the scientific endeavor. In addition, the mentor coaches students as they practice these scientific reasoning skills.

A teaching strategy of this kind has the potential both for developing a canonical knowledge base and for developing competence in the use of the intellectual skills of science. By illuminating the weaknesses of personal theories and providing opportunities for the reconciliation of personal theories with canonical ones, the strategy contributes to the development of the scientific knowledge base. An even more significant value of the method is that it provides students with opportunities to exercise and develop important intellectual skills.

Proposals for teaching science in this way are criticized because they are so time-consuming. Teachers complain that employing such methods would prevent them from covering all the material. Evidence from educational research studies conducted by Benjamin Bloom (Bloom, 1974) many years ago at the University of Chicago suggests that there is no basis for this concern. Because scientific knowledge is cumulative, developing deep conceptual understanding of topics early in a course or program can accelerate learning of topics that follow. An example from physics illustrates this point. Classical mechanics, usually the first topic in beginning physics courses, involves the concept of gravitational potential energy. Typically the second topic in the beginning course is electricity. Electricity is introduced with a water-flow analogy that is based on the correspondence of gravitational and electrical potential energy. However, since students don't develop good understanding of gravitational potential energy from studying classical mechanics, they do not understand the analogy. Consequently, the analogy is misapplied, and considerable time must be expended in the reteaching of energy principles.

Another criticism of the proposed method is that students cannot always plan on learning with a mentor and a support group. This criticism is valid. However, the skills practiced in a social setting under the proper conditions can be internalized. Rather than overtly challenging another person, one can use the internalized skills to challenge an argument or information presented in text. In this sense, the intellectual skills become learning-to-learn-science skills.

Improving science achievement of America's youth requires developing teaching strategies that will facilitate the evolution of personal theories into a canonical knowledge base while developing the intellectual skills that enable further science learning. The proposed teaching strategy addresses both purposes of science teaching while maintaining a correspondence with the conduct of science and taking the nature of human learning into consideration.

REFERENCES

Bloom, B. S. 1974. Time and learning. Amer. Psychol. 29:682-688.

Brumby, M. N. 1984. Misconceptions about the concept of natural selection by medical biology students. Sci. Educ. 68:493-503.

Champagne, A. B., L. E. Klopfer, and J. Anderson. 1980. Factors influencing learning of classical mechanics. Amer. J. Phys. 48:1074-1079.

Cohen, R. A. 1986. A match or not a match: A study of intermediate science teaching materials, pp. 35-60. In A. Champagne and L. Hornig, Eds. This Year in School Science 1986: The Science Curriculum. Washington, D.C.: American Association for the Advancement of Science.

McCloskey, M., A. Caramazza, and B. Greet. 1980. Curvilinear motion in the absence of external forces: Naive beliefs about the motion of objects. Science 210:1139-1141.

Tobias, S. 1986. Peer perspectives on the teaching of science. Change March/April:36-41.

3

Literacy, Numeracy, and Global Ecology

JOHN HARTE

Imagine that it is the third decade of the next century and that your grandchild is starting high-school biology. The average global temperature that year is higher than it has been since 65 million years ago—since the end of the age of the dinosaurs. The Rachel Carsons of the day are warning that because the climate is continuing to warm, plants and animals, including agricultural crops, have to move poleward year after year at the rate of 6 miles per year if they are to remain in their accustomed climate. Because of deforestation in the developing nations, only scattered patches of tropical forest remain, and with the loss of those forests, nearly a quarter of the planet's species have become extinct. In the industrialized nations, over a quarter of the commercial forests have died because of ozone, acid rain, and other air pollutants. Because the stratospheric ozone layer has been depleted by about 20%, more ultraviolet radiation is showering the Earth, resulting in an increase in mutations and cancers; quantitative estimates of the magnitude and consequences of the increased mutation rate are unavailable. And the human population is almost 10 billion, with a yearly

---

John Harte holds a joint professorship in the Energy and Resources Group and the Department of Plant and Soil Biology at the University of California, Berkeley. He is also a senior faculty researcher at the Lawrence Berkeley Laboratory and a senior investigator at the Rocky Mountain Biological Laboratory. He received his undergraduate degree in 1961 from Harvard University and a Ph.D. in theoretical physics in 1965 from the University of Wisconsin. He is the author of two textbooks on environmental science.

increase nearly equal to the number of people living in the United States in 1988.

What would you want that child to be learning in the biology class? More to the point, what should students be learning *today* so that the grim, but entirely plausible, future that I have portrayed does not materialize?

Most current high-school science curricula give, at best, a perfunctory look at already occurring and possible future changes in the biological composition of the planet. Students today are not provided with the knowledge of global-scale processes needed to understand these dramatic changes. They are taught the chemical composition of food, but not the anthropogenic chemical transformations occurring in air, water, and soil that imperil food production. They learn that the growth of microbes in a dish is eventually limited by resources, but remain ignorant of the factors that limit the size of a healthy human population on Earth. They learn the basic idea behind the Krebs cycle, but are not exposed to the fundamental principles that regulate the global biogeochemical cycles and the flows of energy through the biosphere. They learn how information is encoded in DNA, but do not comprehend that our food crops are derived from a small number of wild species and that the future sustainability of food production requires the preservation of genetic diversity on the planet. They emerge from high school thinking that genetic engineering is a panacea, that food comes from a supermarket, that human survival is decoupled from the survival of natural ecosystems, and that clean air and water are luxuries that have to be balanced against economic growth.

I suspect that a common rationale given for such neglect in the high-school biology curriculum is that biology courses should be concerned with "pure" biology, with the *basic* laws, as expressed, for example, in the fundamental architecture of cells, the genetic machinery, and the theory and mechanisms of evolution. All these topics are, of course, important and deserve a prominent place in the biology curriculum. But a dichotomy between pure and applied biology—between puzzles and problems—is unjustified, because at the core of those problems are scientific puzzles that are just as deep and intellectually seductive as are the principles of molecular biology. Indeed, unique biological concepts emerge at the ecosystem or global level and are elucidated by such "applied" fields as conservation biology and the study of the global interconnections among soil, water, air, climate, and life. For example, the discovery of relationships between the area of a habitat and the number of different species that the habitat can sustain resists explanation at the organismic level of analysis. The fact that such relationships have tremendous practical implications for species survival on a fragmented landscape does not diminish their intrinsically fascinating character. Increasing scientific interest in the dependence of human well-being on the maintenance of wild species and natural ecosystem processes is resulting in exciting discoveries at the boundaries of traditional disciplines, such as biology and economics or biology and geology, while study of the stability of the entire planetary life-support apparatus is providing new insight into coevolutionary processes and the dynamics of complex systems.

Knowledge of these system-level phenomena (or "emergent characteristics," as biologists call them) is just as intrinsically fascinating as is knowledge of how the genetic code works . . . and it is as critical to our health as is knowledge of how vitamin C works.

How should global ecology be taught? There will probably be a tendency to place it at the end of the curriculum, atop the basic molecular, cellular, and organismic building blocks laid down at the start of a course. Yet, perhaps it could be taught at the outset. After all, the subject of biology is life on Earth, so is not the study of global biospheric processes a natural place to begin? Of course, such a top-down approach flies in the face of the reductionist philosophy that now dominates biology teaching, and I would not want to claim that there is a compelling argument for either approach today. I would prefer to see each used in different schools, so that a comparison of their effectiveness could be made. It is entirely possible, however, that a top-down approach, stressing at the outset of the course both more natural history and more facts and concepts pertinent to human survival, would motivate students in a way that the traditional curriculum seems unable to do. And it may even mitigate the negative image of scientists that seems to repel students from an interest in the subject. Instead of scientists being viewed as people who invent dangerous things while working in smelly laboratories, perhaps they will be envisioned swimming among endangered whales to study their behavior and hiking up mountains to study acidifying lakes.

I want to make one final point. In the early days of our republic, Thomas Jefferson and others recognized that literacy was essential to the survival of democracy. An illiterate public, they argued, would be preyed on by demagogues and tyrants. Their concerns were taken seriously, and the legacy is that we now attempt to achieve 100% literacy in the United States, although in the time of Jefferson that goal must have seemed quite difficult to reach.

Now, 200 years later, there is yet another vulnerability in our democratic system. Today we confront the threats of uncontrolled technologies capable of destroying the life-support system of the planet. Highly technical testimonies pertinent to the dangers are paraded past Congressional committees and are summarized in the media. Numbers describing megatons, millirads, parts per billion, and kilowatts bombard the public. And experts can be found, or bought, to say almost anything at all on issues affecting our very survival.

As the promotion of literacy was essential to democracy 200 years ago, so the promotion of numeracy is today. A numerate public would not be fazed by very large or small numbers. It would know how to check for consistency between quantitative estimates of risk appearing in the newspapers and what it knows by common sense. A numerate public need not trust the "experts." It need not be bamboozled by those who would numb with numbers and falsely scare or falsely reassure.

How do we teach numeracy? I have been doing it at the university level using a textbook (*Consider a Spherical Cow*, 1988) that I wrote for the purpose. While the text is too difficult for high-school students, the approach I have taken could be adapted to that level. The approach consists of teaching students how to estimate the magnitudes of things and the consequences of events in the world around them, with an emphasis on the use of simple back-of-the-envelope calculation techniques for describing environmental phenomena. The core of the text is a collection of posed and solved word problems that lead the student through the creative process of converting word descriptions of real-world situations into manageable arithmetic. Hundreds of exercises for the reader on a variety of environmental problems are provided as well. While high-school mathematics courses are an appropriate place to teach these techniques, the science courses are where this approach is most critical. I say that with confidence based on my observation that students best retain from their science education the material that they have played creatively with in the courses. So my immodest suggestion is that the pedagogic techniques used in *Consider a Spherical Cow* be adapted to the high-school science curricula.

Global ecology is a fascinating subject. Taught in an imaginative yet applicable way in high schools, it will inspire a new generation of citizens who would be equipped to deal as voters, and perhaps as scientists, with some of the most formidable planetary problems that threaten our survival. We and our children ignore this subject at our peril.

REFERENCE

Harte, J. 1988. Consider a Spherical Cow. Mill Valley, Calif.: University Science Books.

4

"All Is for the Best in the Best of Possible Worlds."

ARCHIE E. LAPOINTE

I selected the title of my paper from Voltaire's *Candide*, because our distinguished chair has eloquently given us two challenges—to be optimistic and to be realistic in our deliberations. I heartily agree with both admonitions. My remarks will be less subtle than Voltaire's satire, but I hope more useful as you ponder a situation that can be viewed either as a disaster or as a unique opportunity. Searching for some cosmic plan in the disarray of today's biology education is probably as fruitless as was Candide's eighteenth-century pursuit. Instead, I'm going to suggest that we approach the problem with some modern management techniques of data assembly, option identification, and decision-making that may move the process along.

I would like to draw on the New Testament, a training manual of the McDonald's Corporation, and Stephen Hawking's best-seller, *A Brief History of Time* (1988), to support my suggestions. At the end of all this, I suspect our chair, like the wise judge she is, will suggest that you disregard my testimony as irrelevant and inappropriate. However, if I'm lucky, I, like a clever prosecutor, will know that some of these comments will linger in the back of your minds and will affect your conclusions.

Archie E. Lapointe is executive director of the Center for the Assessment of Educational Progress. He has been both a teacher and teacher trainer at Louisiana State University and Rutgers University. He was general manager, CTB (McGraw-Hill); vice president, Science Research Associates (IBM); and president, National Institute for Work and Learning.

First some facts:

- There are, right now in 1988, 3,000,000 14-year-olds, most of whom are taking biology.
- Next year, 1989, there will be 3,000,000 more, as there will be each year thereafter.
- National Assessment of Educational Progress (NAEP) data suggest that only about 40% of them are sufficiently prepared intellectually to learn biology.
- The good news, perhaps, is that 80% say they "like" science.
- Fifty percent feel that science is generally useful, will help them in their lives, or will contribute to solving the world's problems of pollution, energy, and food supply.
- Three-fourths of them feel that science will find cures for disease.
- Over half seldom or never look forward to science class.
- Over one-third always or often find their science classes boring, and another 40% agree that "sometimes" it is.
- Of all the 20,000 biology teachers out there, probably 10% are outstanding, 20% are about to retire, and 70% are adequate or better.

These students and these teachers are the very best we have available this year and next. Neither you nor I can change that, so this must be "the best of possible worlds."

May I indelicately stress that these and other realities should be ever present in our thinking. This distinguished group will spend a good bit of energy considering what might be accomplished if there were more time in the curriculum, better-trained teachers, more laboratory equipment, and computers in the schools. The math teachers and the social scientists are doing the same kinds of fantasizing. There won't be any of those improvements in the next 2 years, although unquestionably you have the responsibility to press for them for the future.

Let me suggest another responsibility you should insist on reserving for yourselves and your biologist colleagues. It is you who must set the directions, the objectives, and the content of the biology curriculum. You should come to conclusions quickly, state them clearly, and proclaim them boldly.

NAEP and other test results can tell you what our young people know and don't know, but you must decide whether the news is good or bad.

For example, we know from NAEP results that fewer than 30% of our eleventh-graders in 1986 knew:

- That fats are a source of energy.
- The meaning of "half-life."
- That mosquitoes could develop immunities to pesticides.
- That a diet of french fries and soda lacked nitrogen.

- That data show a higher incidence of disease among smokers.
- That a warmer incubator stimulated seed germination.
- That genetic engineering was being discussed.

On the positive side, 70% or more correctly responded to eleventh-grade questions about:

- Swimming being an inherited behavior.
- The purpose of the doctor's cotton-tipped stick for a sore throat.
- The accumulation of insect poisons in the food chain.
- The claim of an advertisement for a pain-relief product.
- The danger of carbon monoxide.
- The characteristics of junk food.

These are the facts, you must decide what they mean.

Am I pessimistic? No!

Over the last 20 years, NAEP findings in several curriculum areas—in reading, writing, and mathematics, as well as in science—demonstrate rather conclusively that student performance can and does improve. There has been dramatic growth:

- In the basic (as opposed to higher-order) skills.
- Among students attending disadvantaged schools.
- Among black and Hispanic students in reading and mathematics.

These are all school areas where nationally we have clearly set some understandable goals. We went "back to basics," we focused on "minimal competence" in reading and writing, and we funded efforts to address the problems of "at-risk" minority children. These clear and generally accepted goals have been achieved.

I'm optimistic that with equally specific and consistently articulated goals in biology, we could see positive change in performance in the next 5 years.

THE CONTENT

Once you decide the content that should be taught, how do you make sure that the 3,000,000 students in 1992 will be exposed to it? How can you be certain that those 20,000 teachers will "cover" it?

Let's examine the tried and true.

- There are the syllabi. Some schools and school districts will pay attention and will consult them as they select textbooks and write lesson plans.

- There are the "materials of instruction," primarily textbooks. It is alleged and widely accepted that these determine the curriculum in a majority of schools in the country. As a former publisher of textbooks, I have many colleagues in the business who desperately wish this were the case.
- And there are the tests. If these are the so-called *high-stakes* instruments—for example, college entrance tests, advanced-placement tests, or high-school graduation requirements—their content has affected school curricula for decades. This is in spite of the fact that the intention persists that tests should *measure* what is being *taught*.

In some ideal world or closed system, one can imagine syllabi that define the curriculum, structure the textbooks, and dictate test content. Most of us have visited educational systems that are convinced that their own schools are following such an "efficient" design. France and several eastern European countries are good examples of attempts to organize educational environments. Some of our states are more prescriptive than others in these matters—with various degrees of success.

Why doesn't this idea work here in our marvelous but messy society? Sixteen thousand independent school districts and 50 states jealous of their prerogatives are probably part of the reason. May I suggest, however, that what these political entities want as much as their independence is good education for their kids. There are ways to help them achieve the second objective without trampling the first, but imposing a federal syllabus or a national test is not one.

I would urge you not to think of your task as simply developing a plan, setting goals, or defining a curriculum. Your mission, should you choose to accept it, is to improve the biology learning of 3,000,000 14-year-olds a year. It's not impossible! Remember the words of Dan Quayle's grandmother: "You can do anything you put your mind to!"

Our chair's comments remind us of the dangers of narrowly defining objectives as in the Biological Sciences Curriculum Study experience when developers and publishers worked at odds rather than together. She also reminded us of the importance of parental and societal interest in supporting the value of biology education.

Our work at NAEP has convinced me that a broader perspective can be helpful in achieving even narrow specific goals. Five years ago, we redefined NAEP as an information system, rather than a research or testing project. We also recognized that educational policy-makers, as well as the 16,000 school superintendents, in the United States are viewers of the "Today" show and readers of *USA Today*. We then enlisted the help of Carl Sagan and Barbara Walters, whose self-interests are very different, but paralleled ours for brief but effective periods.

SPECIFICS

Point 1

The message must be clear and easily understood. As a disseminator of ideas and information, I marvel at the success of Stephen Hawking's best-selling summary of the mysteries of the cosmos. The rationale for his popularizing effort he lists in his conclusion:

> [However,] if we do discover a complete theory, it should in time be understandable in broad principle by everyone, not just a few scientists. Then we shall all, philosophers, scientists, and just ordinary people, be able to take part in the discussion of the question of why it is that we and the universe exist. If we find the answer to that, it would be the ultimate triumph of human reason—for then we would know the mind of God [Hawking, 1988, p. 175].

It seems to me that your audience is as broad as that and for equally lofty reasons.

What American young people know or don't know about biology affects their lives in real or tragic ways. The opportunities to do harm to themselves abound in their daily lives. The effectiveness of the counseling and education programs designed for them depends on students' understanding of the basic concepts of your discipline. Each one of your students, 4 or 5 years after leaving your classroom, will be voting on issues of pollution, the greenhouse effect, genetic engineering, and chemical and nuclear warfare and will be making nutritional decisions for the next generation.

This is a simple premise—easily understood—but it must be articulated, advanced, and promoted. We continually hear that we don't "value" science education as much as other cultures do—and we don't. For you to be effective in improving biology education, you must assume part of the responsibility for promoting the importance of the learning of science generally.

Point 2

As you develop your course outlines and content, I would urge you to involve publishers from day 1. They do have expertise in designing effective school products, they do know their market, and they do have talented people. Most of them are serious and honest and seeking objectives very similar to your own. They also have hundreds of professional marketing representatives. Most are former teachers who are in the schools daily, promoting new ideas, convincing teachers to try new techniques, and organizing training sessions. In the United States, they represent the regular conduit for getting the results of research into the schools. I would urge their early involvement.

P.S.: It's not easy. They will suggest and insist on compromise in terms of difficulty level, content load, and consistency of presentation. These publishing rules, born of success and failure, frustrate all creative authors, but some are worth paying attention to. Also, publishers are competitive, they must avoid collusion, and they have to protect their markets. The American Association of Publishers can and should help, and the benefits of possible success, it seems to me, are worth the effort.

Point 3

How can those average teachers handle challenging new material? In the New Testament, St. Luke quotes the Lord as commending the unjust steward by saying that "the children of this world are in their generation wiser than the children of light."

Maybe taking a page from the book of the successful children of this world would be enlightening.

In the detailed training manual for owners of McDonald's franchises, founder Ray Kroc is cited over and over again assuring ordinary, everyday citizens of average intelligence that they can become successful owner-operators of Golden Arches outlets. "Success" is measured in financial terms and means an income of $200,000 or $300,000 per year per store. It is achieved in over 80% of the cases.

The training of these ordinary mortals is clear and specific. The goals are limited in number and reinforced consistently. The steps to follow in achieving those targets are spelled out in detail, and deviation from successful practice is discouraged. Quality product, quality service, and consistency are stressed as essential elements of success.

Year in and year out, thousands of franchise managers successfully master the fairly diverse and complex tasks of directing a retail outlet. They hire and train staff, motivate young people, maintain buildings, manage expensive inventories, enlarge their businesses, and engage in community promotional activities. These self-selected lay people are trained to perform all these tasks successfully in very short periods (a few weeks) at places of learning intriguingly called Hamburger Universities.

They are not taught to be great chefs or to be terribly creative, nor encouraged to be innovative. They do provide fast service and consistent quality in a clean, pleasant environment. It's basic, but it's good and perhaps slightly better than functional.

It's almost sacrilegious to advance this model as an option for consideration in the efficient retraining of teachers, who must provide, if not inspired teaching, a least competent, enthusiastic instruction.

We know that the fast-food corporations have been consistent supporters of education and would want to do more to help. We know that this kind of training can be provided quickly and efficiently. We know that it can be effective with most reasonably competent individuals. We know that it falls short of the ideal. Could it be a step on the road?

Point 4

Last year, the National Geographic Society funded an assessment of geography knowledge and skills in order to call attention to the low level of such competence among our young people. The Society's hope is that the availability of data to the broad American public will fuel the debate about the study of geography and reinvigorate interest in the subject.

I can easily imagine an NAEP report in 1992 that would detail the gaps in young Americans' knowledge about biology in ways that would help to form a consensus on the importance of specific aspects of the discipline. NAEP, because of the anonymity of its results (it doesn't identify students, schools, or districts), has a unique ability to focus on issues, rather than individuals, in ways that encourage objective, dispassionate debate. The use of NAEP or the College Board Achievement Tests to call attention to the need might indeed be an intelligent part of your strategy.

Here is an illustration of how NAEP's recent science results are being used with educators, legislators, and business people. The data tend to concentrate lay and professional minds on just what criteria—what standards—are appropriate in today's technological environment.

To communicate the significance of test results to nontechnical minds, NAEP has created scales from 100 to 500 in several curriculum areas. Certain points (levels) on each scale have been identified, labeled, and defined. Each level can be described in terms of what people know and can do if they can function successfully at that level. Examples of tasks that illustrate the knowledge and skills represented by that level are also provided. The defined levels from the NAEP science scale and brief descriptions are listed below.

Level 150	Knows everyday science facts.
Level 200	Understands simple scientific principles.
Level 250	Applies basic scientific information.
Level 300	Analyzes scientific procedures and data.
Level 350	Integrates specialized scientific information.

Level 250, "Applies basic scientific information," can be defined as the ability to interpret data from simple tables and to make inferences about the outcomes of experimental procedures. People at this level of competence exhibit knowledge and understanding of the life sciences and demonstrate some knowledge of basic information in the physical sciences.

TABLE 1 Percentages of 9-, 13-, and 17-Year-Olds at or Above Three Proficiency Levels

Proficiency Level	Age	1977	1986
Level 150 (knows everyday science facts)	9	93.6%	96.3%
Level 250 (applies basic scientific information)	13	49.2	53.4
Level 350 (integrates specialized scientific information)	17	8.5	7.5

In other words:

- From a simple chart they can determine that, with relation to a field mouse, a fox is a predator.
- They can identify the purpose of an experiment that measures plant growth in different types of soils.
- They know that diabetes cannot be transmitted by simple contact.

Most people agree that these are not terribly challenging tasks. There remains, however, the question of what percentage of our 13-year-olds or 17-year-olds should be able to perform at this level?

In the past, simple trend information seemed adequate—that is, we were getting better or falling behind. Increasingly, as recognition grows of the specific requirements for success in a technological society or a world economic competition, our audiences are demanding more precise descriptions of what percentages of young Americans can perform which skills.

Table 1 is a more complete picture of some of the results discussed in NAEP's latest report card, issued on September 22, 1988 (Mullis and Jenkins, 1988).

Is it good enough that 53.4% of our 13-year-olds can apply basic scientific information as described above? That is better than the situation was 10 years ago, when only 49.2% were achieving at this level.

Is it O.K. that 1 of 2 of our 13-year-olds—who are typically in the eighth grade, often experimenting with drugs and alcohol, sometimes pregnant, and usually thinking about careers—can't understand simple biological or chemical concepts or interpret data tables?

More specifically, does this mean that half our 13-year-olds are not prepared to have a positive or satisfying experience during ninth-grade biology? Is it O.K. that the pool of students from which future scientists, engineers, doctors, and nurses will be selected is being turned off or poorly motivated at age 13?

The public and the policy-makers understand issues framed in these terms. Although frustrated by the perennial question, they are open to suggestions and seeking leadership.

I would encourage you to include a testing component in your strategy and formally invite you to take advantage of NAEP's experience and staff, if you decide that either can be of service to you.

REFERENCES

Hawking, S. W. 1988. A Brief History of Time: From the Big Bang to Black Holes. New York: Bantam Books.

Mullis, I. V. S., and L. B. Jenkins. 1988. The Science Report Card: Elements of Risk and Recovery. Princeton, N.J.: Educational Testing Service.

5
The Scientific Revolution in Medicine: Implications for Teachers of High-School Biology

JANET D. ROWLEY

I would like to consider the questions that Dr. Handler has raised: What facts, skills, and perspectives do we wish our students to acquire in the biology classroom? How can we overcome the failure of the public to acknowledge the realities of biology and the processes of nature in the formation of public policy?

ONE APPROACH TO TEACHING BIOLOGY

I have given considerable thought to these problems on a somewhat different level. The University of Chicago has the Benton Foundation Fellowship program for broadcast journalists who come to the campus for a 6-month period of learning and reassessment of their professional role. It has been my task for the last several years to represent the biological sciences. I try to present, in two 2-hour sessions, the fascinating and changing world of molecular genetics as it applies to human disease. I select several items from newspapers reporting on scientific discoveries and give them to the students, as well as the original scientific articles. I try to use these articles to provide the fellows with some, admittedly superficial,

---

Janet D. Rowley received her undergraduate and medical-school education at the University of Chicago. She has studied genetic changes, as measured by chromosomal aberrations in human cancer cells. She has shown that recurring chromosomal changes, especially translocations, are specifically associated with particular subtypes of leukemia and lymphoma.

understanding of DNA technology and how it has revolutionized all aspects of our current understanding of many diseases. I take the view that they cannot afford *not* to understand these important concepts, either in their role as journalists and therefore as educators of the public, or as members of the public themselves. Every year, I use a different example, usually based on several articles that have been published within a few months preceding the course. Two years ago, it was the mapping of cystic fibrosis to chromosome 7; last year, it was the linkage of genes for Alzheimer's disease with DNA probes on chromosome 21. This year, it will be the linkage of colon cancer with DNA markers on chromosome 5 and the growing evidence linking sequential genetic changes in cells with progression of the malignant phenotype.

We are living in a Golden Age of Biology. Virtually every area in this scientific discipline is flourishing as never before. We have the tools to study so many problems, and we have the insights to ask the right questions. The answers to the initial, usually simple questions immediately raise a host of new questions, and so we proceed in an ever-expanding, more sophisticated quest for an understanding of basic biologic processes.

As a physician turned investigator in leukemia who uses genetic analysis to seek an understanding of this fearsome disease, I am awed by the remarkable progress we have made in understanding human disease. None of us can escape the impact that our genes have on us. Some genetic defects, such as "color blindness," are relatively trivial; others—including hemophilia, Duchenne muscular dystrophy, cystic fibrosis, and Tay-Sachs disease—can be devastating.

My approach to the teaching of biology is colored by this background. I believe that the teaching of biology should begin well before high school. It should be centered on human biology; we should take advantage of children's curiosity about themselves to capture their interest and to make them want to learn. I understand that this requires dedicated and well-trained teachers, and you will hear from one such teacher, Frances Vandervoort, here. It also requires clearly written, well-illustrated, up-to-date textbooks. To achieve this goal, such things as loose-leaf texts that can be changed yearly and increased use of video tapes would be important for teaching of basic principles and current applications. These are essential components of a successful educational program; but, unfortunately, they are not sufficient. We need to develop a nationwide recognition, at every level of government, that our strongest national defense is an adequately educated public. Our nation is in far more danger of losing its privileged position because we cannot compete successfully in the world marketplace than because we will be defeated on the battlefield. As we try to redirect our political priorities, we educators must have a solid, but exciting and creative program ready for implementation.

To return more specifically to the topic at hand, I believe that the biology curriculum should concentrate on fundamental principles. The examples should illustrate these principles, but the most important goal should be to impart a basic understanding that can then be applied to a host of similar biological problems. For example, there is very strong evidence that, in some patients, an inherited, therefore genetic, susceptibility is an important predisposing factor in the development of both malignant and nonmalignant diseases. It is now possible in many families to distinguish between individuals who are at risk and those not at risk. How did this come about? The story of this discovery provides a forum for describing principles, as well as specific examples.

HUMAN DISEASES AS EXAMPLES IN BIOLOGY

Let me illustrate the goal of achieving a basic understanding of biological principles by going back to my major premise—that there are so many exciting discoveries in medicine today that you can use them to illustrate any principle you wish to teach. I will pick just a limited area, one that I know something about, namely, the molecular analysis of human genes. Let us take colon cancer, which is of most concern to older individuals, who are at the greatest risk—and who, of course, left high school long ago. These older people are grandmothers and grandfathers or great-aunts and great-uncles; thus, most children know someone or know of someone who has this disease.

DNA as Carrier of Genetic Information

A discussion of colon cancer provides us with a reason to discuss DNA as the carrier of information about how and when cells are to perform certain functions and to explain the notion that this information is contained in discrete units called genes. Some genes are defective before birth, and children who have such genes are born with malformations or with cells and tissues that do not function in the normal way. The ill effects of other genes become apparent only later in life. For those of us who inherit a predisposition to certain malignant diseases, it is possible to find the location of the responsible genes using modern techniques. The basis for these statements is reviewed in McKusisk (1988).

Use of Enzymes for Study of DNA

The next concept required for an understanding of genetic analysis is that DNA can be cut in quite specific places by enzymes that recognize the pattern of the elements making up DNA (Alberts et al., in press). The pattern or sequence of these elements in a particular gene (I am referring here to the nucleotides or DNA bases) may be the same in many different individuals. The DNA from these individuals, when cut into pieces by a particular enzyme and placed in a gelatin slab in an electric current, will give a fragment of identical size when probed with the appropriate gene. Other individuals may have differences in the sequence of DNA bases that are unimportant for gene function, and this may lead to gain or loss of the specific sites cut by the same enzyme. This results in changes in the size of the DNA fragment when it is subjected to an electric current in the gelatin. These changes are DNA polymorphisms, called restriction-fragment-length polymorphisms (RFLPs, or riflips), and they are the basis for much of modern genetic-linkage analysis, especially in humans.

Genetic Linkage

The next concept is that of the linkage of genes and the linkage of DNA probes with genes in cases in which we have not yet identified or cloned the critical gene itself. In fact, this is the situation for many diseases which have been linked with DNA sequences or genes. One can establish the association of specific polymorphisms with a disease in a particular family and then analyze the DNA from a particular individual, to determine the likelihood that the individual is at risk for the disease.

Genetic Analysis of Colon Cancer

I will use the recent studies on colon cancer to illustrate the principles I have just described and their application. As a cytogeneticist, I am especially pleased, as I describe this research, to point out that the initial clue to the chromosomal location of one of these genes came from the study of the chromosomes of a patient with a rare disease that predisposes to colon cancer. This patient had a deletion involving the long arm of chromosome 5, and he had familial polyposis. A report describing this patient was published by Herrera et al. (1986). Ray White and his colleagues in Salt Lake City (Leppert et al., 1987) and Walter Bodmer and his associates in London (1987) recognized the potential usefulness of this information, because the location of the other cancer-related genes had already been identified through their association with specific chromosomal abnormalities. To determine whether familial polyposis was associated with the abnormality of chromosome 5, both groups used pieces of DNA that were known to be polymorphic and that were mapped to this region of chromosome 5. Then they asked, "Are any of these DNA markers linked to the gene for polyposis in families in which a number of individuals in several generations had colon cancer and from whom DNA was available for analysis?" The answer was yes; at least one DNA marker was closely linked to familial polyposis.

The next step was to look at DNA obtained from colon cancers in the general population. Ellen Solomon, an associate of Walter Bodmer, and co-workers (1987) showed by using the same marker probe that the tumor cells in up to 40% of colon cancers had a loss of genes on chromosome 5. These results have been confirmed by a recent study. This study was a collaborative effort of Bert Vogelstein at Johns Hopkins Medical Center in Baltimore, Ray White in Salt Lake City, and Johannes Bos in the Netherlands and their colleagues (Vogelstein et al., 1988). This illustrates the increasing complexity of research, which requires the collaboration of scientists with a variety of skills, often on different continents. Their report describes a complex analysis of 172 colorectal tumor specimens, including those that were premalignant, as well as frank cancers. The different laboratories used DNA probes for genes on three chromosomes, 5, 17, and 18; they also analyzed tumors for mutations in one of the cancer gene or proto-oncogene families, namely, the *RAS* genes. They observed the loss of genes from one or several chromosomes in 25-50% of all the tumors (adenomas and carcinomas) studied. Forty percent of all tumors had a mutation in a *RAS* gene. Their most important observation was that there is a correlation between the degree of malignancy and the number of genetic (usually chromosomal) changes in the cells. Thus, at least one genetic change was detected in only about 25% of very small polyps, compared with 92% of carcinomas. These data provide evidence that the DNA changes that were monitored in this study are likely to be important ones, each of which contributes to a more malignant and more aggressive phenotype.

We know from experimental studies that several changes are required in different genes for a normal cell to change to a fully malignant one. The data in this colon-cancer study show that at least four genes can contribute to the development of a cancer cell. It is quite likely that additional genes will be identified in the future. In this study, the investigators found evidence of a sequence of changes, but it was not an invariant sequence. Thus, when they were identified at all, *RAS* gene mutations and deletion of chromosome 5 occurred during an early, less-malignant stage, whereas a deletion of chromosome 18 followed later, and deletion of chromosome 17 later still. In some patients, deletions were detected only in the middle of the affected chromosome. Mapping the region of deletion provided information on the probable location of the important gene on each chromosome.

These studies on colon cancer are more sophisticated than those reported for lung or breast cancer, because multiple DNA changes in pairs of tumor and normal tissues from the same patient were analyzed. This is just an example of studies that will be described over the next decade. Certainly, future investigations will be even more complex.

As I have already indicated, similar types of analyses are in progress covering a wide range of inherited human diseases, both diseases that result from a mutation in a single gene (for example, cystic fibrosis or sickle-cell anemia) and diseases that result from the interaction of several genes (such as coronary arterial disease or stroke). If American citizens are to comprehend how they can apply this new information to themselves or to their families, they must have an adequate education in biology.

THE HUMAN GENOME MAPPING PROJECT

I have not touched on another compelling reason for emphasizing genetics in teaching biology. I am referring to mapping and sequencing the human genome, which will be a major commitment in biology for the next 2 decades (National Research Council, 1988). For biology, this project is comparable to our space program or to our efforts in high-energy physics. Its cost over this period is estimated to be greater than $3 billion, $200,000,000 a year for 15 years. It would be very helpful if the public were sufficiently educated to understand the benefits of such a commitment. In a time of increasingly limited resources, hard choices must be made. Will members of the public support the level of funding required for successful mapping and sequencing of the human genome if they cannot appreciate its value to them and their families?

The report of the National Reseach Council committee stressed that this project would "greatly enhance progress in human biology and medicine." Although the technology for accomplishing this immense task in an efficient and cost-effective manner is not yet available, the committee's recommendations are to develop a more complete physical map of the chromosomes; then to proceed with sequencing of genes that are functioning, that are expressed in cells; and finally to sequence the pieces of DNA that are between these genes. You will recognize that keeping track of 3 billion nucleotides is a major data management problem that will require substantial improvements in computers and computer programs. This will become increasingly essential as scientists wish to compare different genes to learn more about the correlation between the DNA sequence of a gene and its functional components. Moreover, it has been proposed that parallel projects to sequence the genomes of other species—mouse, Drosophila, etc.—be undertaken at the same time. This will allow scientists to compare the DNA sequences, but perhaps more importantly the organization of genes for the same protein in different species, to achieve an increased understanding of the relationship between the structure of a gene and its function. This information will also provide additional insights into the changes that occur with evolution. Again, a major increase in computer capabilities will be required to make these comparisons in an efficient and effective manner.

Of course, there is concern about the social, legal, and ethical implications of such a project. It is recognized that this project "could provide a great deal of new knowledge about the genetic basis of human disease. However, the effects of that knowledge will be highly colored by the ways its practical implications are interpreted" (National Research Council, 1988, p. 101).

CONCLUSION

I have tried to give examples of the progress being made in medicine today and to show how the teaching of a few general principles can provide a framework for students to understand many of the new discoveries in genetics. It will not be easy to help students achieve the necessary level of such an understanding. However, I believe that they can appreciate the importance of this knowledge and that this appreciation, provided by enthusiastic teachers and first-rate instructional material, will lead to a better-educated and more-informed American public.

REFERENCES

Alberts, B., D. Bray, J. Lewis, M. Raff, K. Roberts, and J. D. Watson. In press. Molecular Biology of the Cell. 2nd ed. New York: Garland.

Bodmer, W. F., C. J. Bailey, J. Bodmer, H. J. R. Bussey, A. Ellis, P. Gorman, F. C. Lucibello, V. A. Murday, S. H. Rider, P. Scambler, D. Sheer, E. Solomon, and N. K. Spurr. 1987. Localization of the gene for familial adenomatous polyposis on chromosome 5. Nature 328:614-616.

Herrera, L., S. Kakati, L. Gibas, E. Pietrzak, and A. A. Sandberg. 1986. Brief clinical report: Gardner syndrome in a man with an interstitial deletion of 5q. Amer. J. Med. Genet. 25:473-476.

Leppert, M., M. Dobbs, P. Scambler, P. O'Connell, Y. Nakamura, D. Stauffer, S. Woodward, R. Burt, I. Hughes, E. Gardner, M. Lathrop, J. Wasmuth, J.-M. Lalouel, and R. White. 1987. The gene for familial polyposis coli maps to the long arm of chromosome 5. Science 238:1411-1413.

McKusick, V. A. 1988. The Morbid Anatomy of the Human Genome: A Review of Gene Mapping in Clinical Medicine. Bethesda, Md.: Howard Hughes Medical Institute.

National Research Council. 1988. Mapping and Sequencing the Human Genome. Washington: D.C.: National Academy Press.

Solomon E., R. Voss, V. Hall, W. F. Bodmer, J. R. Jass, A. J. Jeffreys, F. C. Lucibello, I. Patel, and S. H. Rider. 1987. Chromosome 5 allele loss in human colorectal carcinomas. Nature 328:616-619.

Vogelstein, B., E. R. Fearon, S. R. Hamilton, S. E. Kern, A. C. Preisinger, M. Leppert, Y. Nakamura, R. White, A. M. M. Smits, and J. L. Bos. 1988. Genetic alterations during colorectal-tumor development. New Engl. J. Med. 319:525-532.

6

High-School Biology Training: A Prospective Employer's View

HARVEY S. SADOW

INTRODUCTION: THE PROBLEM

I do not teach biology at the high-school or any other level, nor do I now have a certificate to teach anything, including biology. I have not engaged in biological research for roughly 20 years. I am certainly not a specialist in, nor even more than perhaps modestly informed about, curriculum in high-school biology. Finally, my days as an educator are so far in the dim and distant past that I really cannot claim more than "having been. . . ." Thus, having completely destroyed my credibility by acknowledging my lack of credentials, I will demonstrate my temerity by talking about high-school biology education today, but especially today in the face of tomorrow's needs, as an employer of a large body of research scientists, physicians, and technicians without advanced or collegiate education.

You may justifiably ask why I am here, having obviously admitted my limitations; to that the answer must be that I have a concern about the teaching of the scientific disciplines, such as biology, in our high-school programs. I am compelled, however, in that concern by the recognition of

---

Harvey S. Sadow is chairman of the board of Boehringer Ingelheim Corporation and its former chief executive officer and president. He is a member of the board of the Pharmaceutical Manufacturers Association and chairman of the board of the Pharmaceutical Manufacturers Association Foundation. Dr. Sadow is also president of the Connecticut Academy of Science and Engineering. He received a B.S. from the Virginia Military Institute, where he serves on the board of visitors; an M.S. from the University of Kansas; and a Ph.D. from the University of Connecticut.

another trend that forces the issue. The United States, for many reasons, has passed rapidly in the last 2 decades from a pre-eminently manufacturing economy to one of service. If the United States is to regain its pre-eminent position in the production-technological areas, it must commit itself to enhanced scientific innovation, which, of course, means the stimulation of the evolution, and conversion to practice, of new ideas. As has been said about the manufacturing economies of many states, including my own Connecticut, in a changing, competitive world, it is necessary to innovate or die—at least on the economic limb!

Another fact is increasingly inescapable, and it is brought home daily in our experience in western Connecticut, where the company I have led is. There is a significant and growing shortage of technically qualified or even trainable labor, which seriously threatens the innovative high-technology R&D and manufacturing components of our company.

Dr. Handler, in her opening remarks, cited the observations of Armstrong and co-workers (the Education Commission of the States) concerning the relatively poor American student achievement in scientific education, compared with that of other developed countries, emphasizing that science instruction has had a low priority; the teachers of science are inadequately trained; there are teacher shortages in certain basic scientific fields, accompanied by a decline in the enrollment of high-school students in science courses and, among other things, the lack even of a consensus as to why science should be taught, what should be taught, and to whom, and thus, how the process can be changed. Perhaps even more troubling than the reference to Armstrong et al. was the statement that these young people are deficient in their understanding of biology as a "coherent discipline." Reference has been made to both public and political failure to acknowledge, or perhaps even create public policy concerning, educational realities, as in the field of biology. Then again, American mores and attitudes have changed over the years since the end of World War II. Discipline, especially self-discipline, seems to have evaporated in the process of developing young people. Is it any wonder that the undisciplined would, of necessity, seek to avoid the strict disciplines of either the physical or the natural sciences, especially if there are easier ways to get high-school diplomas? The problem, therefore, of attracting the interest of these young minds to the field of biology, and keeping it, is one of the reasons for this conference.

SHOULD BIOLOGY BE TAUGHT IN HIGH SCHOOL?

The answer for me is unequivocally "Yes!" Biology is no longer simply a descriptive field in the range of the natural sciences. It has, just in the last 10-20 years, changed to a vibrant, dynamic multidiscipline, which has invaded chemistry, physics, mathematics, and indeed even the technologies of engineering, especially electronics. It seems to me that the important subordinate questions suggested by Dr. Handler, which must also be asked, include: "To Whom?" "What?" And perhaps even precedent to these questions, "Why?" I will try, from the prospective employer's point of view, to answer.

WHO SHALL BE TAUGHT?—AND WHY?

Young minds—if they are to benefit from the explosion of new information, which will certainly, in some way, touch everyone's life—must be prepared to adapt, early on, to the present dynamism of biology. That dynamism, of necessity, directly influences biology education. That adaptive preparation must be based on the soundest possible foundation of basic knowledge and understanding of biology as the basic science of life itself.

I believe that today, in most high schools, there is at least one required course in "general science." This affords an initial exposure, however superficial, to very basic information on the nature of living things. Obviously (at least to me), it would be preferable to offer a basic course in biology as a scientific discipline to all whose interest in the field may have been stimulated either by such a basic science course or, if none were available, by reading, by advice from career guidance counselors, or by completion of courses, particularly in basic chemistry or physics. Of course, prior basic knowledge in physics and chemistry would be highly desirable to ensure a better understanding of the processes and mechanisms prevailing in living organisms.

To those young people who may be college-bound, I would "sell" the virtue of the study of basic biology, as well as chemistry and physics, as an assurance of doing better, earlier, in the college-level study of these sciences. To those students not headed for college who show any aptitude for the scientific disciplines, I would also "sell" the study of biology as fundamental job preparation, especially for technician jobs. Even if the student shows no aptitude for biology as a scientific discipline, study of the subject might still be encouraged, if only for the awareness and understanding it can afford of basic life processes seen or experienced day by day throughout one's life.

Even though the interests and goals of high-school students are not all the same, it should be possible to bring home the fact that in the study of biology, there is something for everyone.

WHAT SHOULD BE TAUGHT?

Now, the answers get a bit stickier. What will be taught depends on who will be taught. In a sense, we are dealing with divergent populations: the college-bound, including those who will seek only undergraduate degrees, with or without a major in biology or any other scientific discipline, and those who will ultimately pursue biological science-related professional degrees and careers; and the non-college-bound, whose exposure, if any, to biology as an academic pursuit will be an isolated or terminal one and who may or may not find jobs in biological science-related fields possible, but who, if they do, will receive further on-the-job technical training in industry, clinical laboratories, or other workplaces.

Should all those divergent student populations be taught the same way? The answer must obviously be "yes." All, regardless of direction of later pursuits, would benefit from a few essential basics in biology education. To my way of thinking and experience, these essentials might include the following:

- An understanding of the structure and function of living organisms; thus, fundamental life processes, regardless of form.
- Application of that understanding of life processes to things seen in the world around us.
- An understanding of the "scientific method" and its application.
- Learning by doing—simple biology laboratory procedures, not only to enhance hands-on experience, but also to develop basic manipulative skills.

These basics, to which I am sure others might be added, should be taught to all high-school students without regard for the post-high-school education or work intentions. For the future college students, they will provide foundations for the next stage of the learning process, as intended. Good and sound curricula taught by motivated and adequately trained teachers should open young minds to the opportunities in the biological sciences, and especially to the value of at least basic biology education and to the appreciation of how things around us are affected by disturbances in the balances of life processes (e.g., environmental pollution, disease, and atmospheric change, to name just a few). High-school biology education can encourage the uncertain student of certain potential to begin to discriminate and thus choose previously unknown or unappreciated further foci in later education and ultimate career pursuit. For the fortunate young person who always knew what he or she wanted to do, in the areas founded on or related to biological sciences, high-school biology educational exposure may prove to be the first real confirmation of the wisdom—or even lack thereof—of that presumption.

Of course, for the student motivated to pursue some career-related interest in biology, additional material, probably closer to applications of the science, might, given the institutional resources, be offered—but in advanced courses. Thus, one could foresee course work in the principles and applications of genetics, as in zoology, botany, biotechnology (DNA manipulation), and environment as a biological entity. The list is much longer and might even include, with caution, societal concerns with biology. However, the issue here might be how much is enough or too much. I say that, because of the evident mismatch between expectations and capacities, both individual and institutional, with which everyone in high-school education must live.

Returning to the view of the issue that I hold as a prospective employer, the college-bound are of less immediate concern in relation to high-school biology training. Except for adequacy of preparation to receive more education in biology, the young person leaving college will, it is hoped, have already gone beyond basics and thus be ready for a position, even if of limited scope or responsibility, in research, development, or related biological technology at the technician or more advanced level.

What about the non-college-bound students? Regardless of the reasons for that decision, whether they are economic or social, let us assume some capacity to learn, absorb, and even apply basic high-school biology training. We have found that with good basic biology education, these youngsters can quickly grasp principles and practice in a typical biochemistry, toxicology, physiology, or even pharmacology research laboratory or biological quality-control or clinical-assay laboratory. The quick absorption and understanding of a technician's work, thanks to high-school biology training, helps to make these young people productive economic contributors to their jobs when receiving on-the-job training. That means earlier advancement and better job opportunities, albeit at technician levels. For some, however, on-the-job training has reinforced interest in biological science as a career; and, family circumstances permitting, it has encouraged at least a few to seek higher education as an assurance of the achievement of greater biology-related career goals.

Observation of weaknesses in high-school biology training for these students usually illuminates two prime areas:

- Inadequate manipulative training and thus limited laboratory procedural skills.
- Little or no real knowledge of scientific methods or their application.

CONCLUSION

Having made these views known, I should say that I recognize that probably everything that I have said here has been said before, many times. As in the educational process itself, however, repetition can lead to recognition, to acceptance, and to ultimate action. Biology, once the "easy" science in high school, and even in our colleges, is now both the foundation and the capstone for some of the greatest advances in our understanding of life processes in health and disease and thus of our capacity to intervene successfully and restore balance. To my mind, therefore, it is our obligation to lay solid foundations of basic knowledge, and thus understanding of life processes, in the high-school setting, so that our young citizens may benefit, as fully as their individual capacities permit, from our progress in this field.

PART II

Objectives of Biology Education and Measurement of Achievement

7
Issues in Objectives and Evaluation

JAMES T. ROBINSON

GOALS, OBJECTIVES, AND OUTCOMES

Biology is taken by most students during the high-school years. It is incumbent on us to re-examine why biology is important for most or all students and what we expect the benefits of biology education to be, both for the individual and for society at large.

Several issues in the field of goals, objectives, and outcomes of biology education will need to be resolved as the Committee on High-School Biology Education addresses its tasks. "Scientific literacy" has been espoused as a social imperative for a society affected so importantly by science and technology. The American Academy of Arts and Sciences (1983) devoted an entire issue of its proceedings to elaborate the meanings of scientific literacy. That same year, in *Educating Americans for the 21st Century* (National Science Foundation, 1983), the National Science Board Commission on Precollege Education in Mathematics, Science and Technology "found that virtually every child can develop an understanding of mathematics, science and technology if appropriately and skillfully introduced at the elementary, middle and secondary levels."

The commission recommended the following criteria for improving high-school science (National Science Foundation, 1983, p. 98):

James T. Robinson is a former executive director for curriculum and evaluation in the Boulder Valley (Colorado) School District. He served as a staff officer for the Biological Sciences Curriculum Study.

- Drastically reduce the number of topics covered in high-school science courses.
- Direct attention toward the integration of the remaining facts, concepts, and principles within each discipline and with other sciences and such areas as mathematics, technology, and the social sciences.
- Select ideas that can be developed honestly at a level comprehensible to high-school students.

- Develop ideas out of experimental evidence that high-school students can gather or, at least, understand.

- Tie ideas into other parts of the course, so that their use can be reinforced by practice.
- Let all courses provide opportunities to develop the ability to read scientific materials.

These criteria raise issues for biology education that pervade all areas of our concern at this conference. If outcomes are to be determiners of curricula and evaluation, then the other subjects of this conference are derivative from the goals, objectives, or outcomes to be formulated as a major function of the committee.

Several questions are proposed for consideration here. Should high-school biology goals and objectives:

- Be designed for all students, or should separate courses be developed for students with different interests and goals?
- Be formulated in the context of a science and contribute to public understanding of science or as a separate discipline independent of other sciences?
- Include the application of knowledge and understandings or be limited to the acquisition of knowledge?
- Include attitudes toward science and technology and developing interest in biology and other sciences?
- Include ethical and societal issues of science, biology, and technology?
- Specify the development of problem-solving, critical thinking, and other "higher-order" thinking skills?
- Be measurable or assessable in some objective and "practical" manner?

The literature is fairly consistent in an affirmation of positive positions on these questions, but in the classrooms in high schools these issues are not settled at all, in stated objectives, actual practice in instruction, or testing and evaluation. Also, coverage of subject matter dominates instruction (Stake and Easley, 1978); it is questionable whether retaining the current breadth of coverage will permit students to attain the other outcomes specified above.

It is ironic that *Educating Americans for the 21st Century* lists drastic reduction of content as a major need in high-school science and then, in the statement of outcomes, includes all the major areas currently included in high-school biology. For example, the National Science Board commission, in discussing science education and high-school biology, proposed that scientific education programs in K-12 should be designed to produce the following outcomes (National Science Foundation, 1983, p. 44):

- Ability to formulate questions about nature and seek answers from observation and interpretation of natural phenomena.
- Capacities for problem-solving and critical thinking in all areas of learning.
- Innovative and creative thinking skills.
- Awareness of the nature and scope of a wide variety of science- and technology-related careers open to students of varied aptitudes and interests.
- Basic academic knowledge necessary for advanced study by students who are likely to pursue science professionally.
- Scientific and technical knowledge needed to fulfill civic responsibilities and improve students' own health, life, and ability to cope with an increasingly technical world.
- Means to judge the worth of articles presenting scientific conclusions.

The commission proposed that general biology in high schools should emphasize biology in a social and ecological context. Biology should enable students to attain the following outcomes (National Science Foundation, 1983, p. 98):

- Understanding of biologically based personal or social problems and issues, such as health, nutrition, environmental management, and human adaptation.
- Ability to resolve problems and issues in a biosocial context involving value or ethical consideration.
- Continued development of students' skills in making careful observations, collecting and analyzing data, thinking logically and critically, and making quantitative and qualitative interpretations.
- Ability to identify sources of reliable information in biology that they may tap long after formal education has ended.
- Understanding of basic biological principles, such as genetics, nutrition, evolution, reproduction of various life forms, structure-function relationships, disease, diversity, integration of life systems, life cycles, and energetics.

The problems associated with formulating goals, objectives, or outcomes are formidable. First, a national consensus on such a statement would be extremely difficult to attain; and second, evidence seems to support the observation that classroom instruction is determined more often by the textbook used by teachers than by statements of goals in curriculum guides (Stake and Easley, 1978, pp. 13:59-64).

The issues implied here have included the question of the target population for high-school biology, its range of content, its context (social, technological, scientific), and its attention to application of knowledge and to the inclusion of higher-order thinking skills. Sorting these issues out is essential and is related to all the other dimensions of high-school biology.

EVALUATION STUDIES

The preliminary report by the International Association for the Evaluation of Educational Achievement (IEA, 1988) presents international comparisons of student achievement. A biology test of 30 items was given to twelfth-graders in 17 countries. Table 1 shows the numbers of items in the various topics.

The U.S. sample taking the biology test was drawn from 43 schools with a total of 659 students taking a second year of high-school biology. There are no U.S. data on first-year biology students, nor for nonscience students. Validity of the biology test was measured by three indexes (IEA,

TABLE 1 Biological Content Areas and Numbers of Items Given to Twelfth-Grade Students in 17 Countries[a]

Biological Topic	No. of Items
Transport and cellular material	2
Concept of gene	6
Diversity of life	1
Metabolism of the organism	3
Regulation of the organism	3
Behavior of the organism	3
Reproduction and development, plants	1
Reproduction and development, animals	2
Human biology	2
Natural environment	1
Evolution	6
Total	30

[a]Five items, undesignated, were cut from the test given to students in the United States.

1988, p. 93): a curriculum-relevant index (0.76), a test-relevant index (1.00), and a curriculum-coverage index (1.00).

Interpreting the results of the IEA biology test cannot be straightforward, because of several conditions. Five items were dropped from the test given to U. S. students, and "the scores (comparing countries) are presented in percentage frequencies but it must be noted that the United States with 25 items is being compared with other countries with 30 items. The reduced number of items in percentage form will result in a reduced range" (IEA, 1988, p. 46). A second year of biology may be inferred to be an advanced course for able students, but in the district in which I recently worked, a second biology course is offered for students who do not want to take chemistry or physics, but wish to take more science. I do not know how prevalent this practice is. However, the biology scores are reported as scores of the "elite" (IEA, 1988, p. 73).

The mean achievement of students in the United States for the 1986 administration was 37.9%, with a K-20 reliability of 0.669, which indicates that the items are not very homogeneous in difficulty. The highest national score reported was for Singapore, with a mean of 66.8%. With the limitations of the test data, the United States had the distinction of having the lowest mean percentage score on the biology test. The next lowest mean percentage was attained by Italy, with a mean of 42.3%.

To give you a flavor of the test, one item asked, "What initially determines whether a human baby is going to be a male or a female?" Response options and percentages of U. S. students selecting them were (IEA, 1988, p. 120):

A. The DNA in the sperm.	48.44%
B. The DNA in the egg.	6.00%
C. The RNA in the sperm.	9.17%
D. The RNA in the egg.	2.72%
E. The DNA and RNA in both sperm and egg.	33.25%
No response.	0.42%

I reviewed *Modern Biology* (Otto and Towle, 1985) and *Biological Sciences: An Ecological Approach* (BSCS, 1982) to find out how they treated the subject. In both books, although they treat the subjects differently, sex inheritance is explained through X and Y chromosomes, and the more extensive presentation of DNA is associated with the function of DNA and RNA in gene action. This linking of DNA and RNA in gene action could have led students to select response E.

The preliminary report of the IEA study will be followed by more detailed analyses of the test data and other variables not currently processed. The main report will be published in 1989.

The National Assessment of Education Progress (Blumberg et al.,

1986) piloted the development and testing of higher-order thinking skills in science and mathematics for potential use in future national assessments. Exercises included hands-on activities of students to solve problems. Three modes of administration were used: intact classes with paper-and-pencil tasks, but with materials as stimuli; station activities with students rotating from station to station, each station having apparatus and investigations; and full investigations administered to individual students with an observer using a checklist to record what students did as they performed an investigation. Third-, seventh-, and eleventh-graders were tested in 12 school districts. In one example of a station problem, eleventh-graders were to examine a set of 11 vertebrae, put them into three groups, and explain the similarities of the bones in each group. Cat, rabbit, and dog vertebrae were used. Fifty-four percent of the students were able to place the thoracic, cervical, and lumbar vertebrae into their proper groups. Another 20% grouped all but the atlas vertebra appropriately. Sixty-seven percent of the students provided at least one distinguishing feature for each group of vertebrae (Blumberg et al., 1986, Part II).

BIOLOGY TEACHERS

Only one recent study was found regarding biology teachers' knowledge of biological concepts. This study was reported in Cleveland, Ohio, newspaper, *The Plain Dealer* (Epstein, 1987), and found that only 12% of biology teachers surveyed correctly defined the modern theory of evolution. This study was based on written responses to items about evolution from 404 Ohio high-school biology teachers, about one-third of the biology-teacher population. Michael Zimmerman, a biology professor at Oberlin College who conducted the study, also found that 37.7% of the teachers surveyed favored teaching creationism and three-fourths felt that creationism was a favorable explanation for the origin of life (Epstein, 1987).

From these two studies and from those reported by other panelists, I believe we can conclude that major reconsideration of the goals and objectives of high-school biology education and of methods of assessing student interests, achievement, and attitudes is important.

EVALUATION IN HIGH-SCHOOL BIOLOGY

Schools and such courses as biology are continuously subjected to informal evaluation by their many publics: parents, students, administrators, teachers, scientists, business men and women, and national groups. These informal evaluations carry great weight about the quality of education in each community and in the country as a whole. Efforts to inform these many judgments by more objectives measures and indicators of student achievement have been low-technology, low-budget items. My judgment here is based on comparison of expenditures for accurate instruments for measurement in astronomy, physics, biology, medicine, and space activities.

As I looked over evaluation instruments for biology, I saw little change in the last 50 years. A few efforts, such as those of the Educational Testing Service (Dressel and Nelson, 1956) and the Biological Sciences Curriculum Study (Schwab, 1963; Klinckmann, 1970; Mayer, 1978), provided teachers with resources for improving multiple-choice test items in biology. These resources provided sample items for going beyond pure recall and enabling students to demonstrate their capabilities of interpreting experimental data, applying knowledge to novel situations, and interpreting graphed data.

More recently, the National Research Council (Raizen and Jones, 1985; Murnane and Raizen, 1988) has broadened the discussion of evaluation to include indicators of quality in science and mathematics education.

The major issues in evaluation revolve around purposes and related instruments. Do we want to sort students on test scores similarly to the way we can sort students on height or weight? If so, we have norm-referenced tests (most standardized tests) that are designed to do just that. Norm-referenced tests are constructed, and items selected, to provide a normal distribution with mean and median at the 50th percentile. Most standardized tests are renormed about every 10 years. The new tests may be more or less difficult than the previously normed tests, but the new norms have statistical characteristics similar to those of the old.

Another characteristic of the commonly used standardized tests is that they are designed to measure general knowledge and are not directly related to what is taught in any particular classroom.

Within the last 20 years, criterion-referenced tests have been developed, especially as part of the "minimal-competence" movement. Criterion- and domain-referenced tests are directly interpretable in terms of a "standard." One problem with these tests is determining what the standard should be, other than in arbitrary ways. A second problem is the desire to make inferences about student competence by generalizing beyond an ability to achieve similar scores on similar paper-and-pencil tests (Haertel, 1985).

This identifies a second issue: "Can a single instrument serve all the purposes desired?" Among the purposes are diagnosis and guiding instruction, rank-ordering students, judging instructional quality, judging curricular quality, forcing curriculum and instruction to move in a particular direction, predicting future performance of individuals, and formulating policies for schools, districts, or states.

Another issue is measuring student performance in a way different from the "recognition knowledge" that is assessed in multiple-choice formats. A great deal of interest is developing in generating alternatives to both the commonly used forms of testing. One such alternative is performance testing: assessments that call on the examinee to demonstrate specific skills and competences and to apply them to novel situations (Stiggins, 1987). Performance assessments have "four basic components: a *reason* for assessment, a particular *performance* to be evaluated, *exercises* to elicit that performance, and systematic *rating* procedures" (Stiggins, 1987, p. 34).

Laboratory work is considered to be an important and necessary means of enabling students to attain the essential goals of biology education, but assessment of any unique contributions of laboratory work is rare (Robinson, 1979). Laboratory practicals have been used, but Gallagher (1987) commented that, despite the prevalence of laboratory work in science, we know very little about its effects on high-school biology achievement in the United States. Indeed, both effective and comprehensive evaluation practices and evaluative instruments are a critical need for the improvement of high-school biology. Tamir and co-workers (Tamir, 1974; Tamir et al., 1982) developed and have placed in use a laboratory practical in the schools of Israel, but evidence of its use outside Israel is lacking.

A science-test review panel convened by the National Research Council (Murnane and Raizen, 1988) carefully examined nine science tests. The panel consisted of 12 scientists and high-school science teachers. They made three recommendations to avoid the misuse of science-test results (Murnane and Raizen, 1988, p. 180):

- Results from tests constructed for one purpose . . . should not be used for a quite different purpose.
- School or classroom average test scores should not be applied to individuals, and individual test scores should not be interpreted as a rating or ranking of the persons, but only of performance on a test that assesses specific skills.
- Test results or tests of the kind reviewed should not be used as the major force driving curriculum and instruction.

CONCLUDING REMARKS

Goals, objectives, and outcomes and the evaluation procedures used to assess them are two critical aspects of any proposal for policy formulation for high-school biology. I did not mention accountability earlier, but the accountability movement has stimulated the development of evaluation processes and can pressure curriculum and instruction to be concerned with only the aspects of biology that can be easily measured. In many instances, especially with many standardized testing programs, the student is forgotten in the process. It would seem that a first criterion of evaluation programs would be that they have significance to the students themselves.

The technology of assessment needs to have infusions of creativity, research, and development. Surely, computers and associated technologies can provide for more useful, instructive, and informative evaluation information. Devising more effective evaluation instruments and procedures requires that we be clear and specific about the purposes of biology education and the outcomes that we can reasonably be expected to attain with the approximately 134 hours we have to help a very diverse group of adolescents attain the understanding we propose.

REFERENCES

American Academy of Arts and Sciences. 1983. Scientific literacy. Daedalus 112:Spring issue.

Blumberg, F., M. Epstein, W. MacDonald, and I. Mullis. 1986. A Pilot Study of Higher-Order Thinking Skills Assessment Techniques in Science and Mathematics. Final Report. Part II. Princeton, N.J.: National Assessment of Educational Progress.

BSCS (Biological Sciences Curriculum Study). 1982. Biological Science: An Ecological Approach. BSCS Green Version. 5th ed. Boston: Houghton Mifflin.

Dressel, P. L., and C. H. Nelson. 1956. Questions and Problems in Science. Test Item Folio No. 1. Princeton, N.J.: Educational Testing Service, Cooperative Test Division.

Epstein, K. C. September 3, 1987. Many Ohio science teachers favor study of creationism. The Plain Dealer. Cleveland, Ohio.

Gallagher, J. J. 1987. A summary of research in science education–1985. Sci. Educ. 71:271-455.

Haertel, E. 1985. Construct validity and criterion-referenced testing. Rev. Educ. Res. 55:23-46.

IEA (International Association for the Evaluation of Education Achievement). 1988. Science Achievement in Seventeen Countries. A Preliminary Report. New York: Pergamon Press.

Klinckmann, E. 1970. Biology Teachers' Handbook. 2nd ed. New York: John Wiley and Sons.

Mayer, W. V. 1978. Biology Teachers' Handbook. 3rd ed. New York: John Wiley and Sons.

Murnane, R. J., and S. A. Raizen. 1988. Improving Indicators of the Quality of Science and Mathematics Education in Grades K-12. Report of the National Research Council Committee on Indicators of Precollege Science and Mathematics Education. Washington, D.C.: National Academy Press.

National Science Foundation. 1983. Educating Americans for the 21st Century: A Plan of Action for Improving Mathematics, Science and Technology Education for All American Elementary and Secondary Students So That Their Achievement Is the Best in the World by 1995. A Report to the American People and the National Science Board. Washington, D.C.: National Science Board Commission on Precollege Education in Mathematics, Science and Technology.

Otto, J. H., and A. Towle. 1985. Modern Biology. New York: Holt, Rinehart and Winston.

Raizen, S. A., and L. V. Jones. 1985. Indicators of Precollege Education in Science and Mathematics. A Preliminary Review. Report of the National Research Council Committee on Indicators of Precollege Science and Mathematics Education. Washington, D.C.: National Academy Press.

Robinson, J. T. 1979. A critical look at grading and evaluation practices. In M. B. Rowe, Ed. What Research Says to the Science Teacher. Vol. 1. Washington, D.C.: National Science Teachers Association.

Schwab, J. J. 1963. Biology Teachers' Handbook. New York: John Wiley and Sons.

Stake, R. E., and J. Easley. 1978. Case Studies in Science Education. Vol. II. Urbana-Champaign, Ill.: Center for Instructional Research and Curriculum Evaluation and Committee on Culture and Cognition, University of Illinois, Urbana-Champaign.

Stiggins, R. J. 1987. Design and Development of Performance Assessments. NCME Instructional Module. Educational Measurement: Issues and Practice. Washington, D.C.: National Council for Measurement in Education.

Tamir, P. 1974. An inquiry oriented laboratory examination. J. Educ. Measure. 11:25-33.

Tamir, P., R. Nussinovitz, and Y. Friedler. 1982. The design and use of a practical tests assessment inventory. J. Biol. Educ. 16:42-50.

8

Assessing Student Understanding of Biological Concepts

CHARLES W. ANDERSON

The students quoted below were juniors and seniors in a nonmajors' biology course at Michigan State University. On the average, they had completed 1.9 years of previous biology courses. The first pair of questions, given in multiple-choice form, concern their ideas about sources of energy for plants and animals.

Questions:

A bean plant needs energy to survive and grow. What is (are) the source(s) of the energy that a bean plant uses?

A human also needs energy to survive and grow. Where do you think that a person gets the energy that he or she needs? (Circle all correct.)

Student responses:

S1: Bean plant: Air, water, sun, soil
Person: Air, water, meat, potatoes
S2: Bean plant: Air, water, sun, soil
Person: Air, water, sun, exercise, meat, potatoes
S3: Bean plant: Air, water, sun, soil
Person: Air, water, meat, potatoes
S4: Bean plant: Water, sun, soil

Charles W. Anderson is an associate professor in the Department of Teacher Education, Michigan State University. His research focuses on teaching for understanding and conceptual change in science learning.

Person: Air, water, sun
S5: Bean plant: Air, water, sun, soil
Person: Air, water, sun, exercise, meat, potatoes

Not surprisingly, given the array of ideas that the students had about sources of energy, they also had a variety of ideas about energy conversion processes, such as photosynthesis:

Question:
How do you think that a biologist would define the term "photosynthesis"?
Student responses:
S1: The conversion of light to energy.
S2: Taking in inorganic material for use in the organism.
S3: Changing sunlight energy into useful energy form.
S4: The process by which a plant obtains energy by turning sunlight into CO_2.
S5: The process by which plants convert the sunlight into needed nutrients.

The following question was designed to assess students' understanding of energy pyramids. Clearly, most students invoked different concepts.

Question:
A remote island in Lake Superior is uninhabited by humans, but supports populations of white-tailed deer and wolves. It is left undisturbed for many years. What will happen to the average size of the populations over time? (Multiple-choice predictions, open-response explanations.)
Student responses:
S1: This question cannot be answered because we have no idea of the amount of deer and wolves on the island and the time.
S2: The deer will all die or be killed because of their white tails. The wolves will find it easy to find them.
S3: On the average, there will be a few more wolves than deer, because the wolves will kill the deer for food.
S4: On the average, there will be many more wolves than deer because wolves are carnivorous and deer would become food source.
S5: On the average there will be many more wolves than deer. Survival of the fittest.

The final example focuses on students' ideas of how the process of evolution occurs.

Question:

Cheetahs (large African cats) are able to run faster than 60 miles per hour when chasing prey. How would a biologist explain *how* the ability to run fast evolved in cheetahs?

Student responses:

S1: The cheetah's ability to run faster may be influenced by skeletal changes over many years. His legs may have become longer and he may be better adapted for running at high speeds because of his need to do so in order to survive.

S2: Since cheetahs are smaller animals they are not very strong compared to other animals such as lions. Since they could not fight off animals effectively enough they have learned to escape their hunter.

S3: The cheetah's running ability changed due to its environment. As they evolved they needed to run faster to catch faster animals.

The tests from which these responses were taken were pretests administered as part of a project to improve instruction in the course (Anderson et al., in press; Bishop and Anderson, in press). At the beginning of this project, I believed that our system of biology education was, if not working perfectly, at least working. The results of the pretests and posttests were discouraging, though, to someone with those beliefs. The tests quoted were taken from the middle of the stack of tests that I still have in my file folders; they are neither particularly better nor particularly worse than the other tests in the stack. The students' level of performance on the pretests was generally low; I saw little evidence of knowledge beyond that which my 12-year-old daughter is picking up from watching nature shows on television. Furthermore, there were no significant correlations between the level of performance and the amount of previous biology coursework that students had taken (the range was from less than 1 to more than 4 years). Those biology courses did not seem to be doing the students much good.

These results led me to the position that I will take and elaborate on in this paper: *Most students are not learning anything useful in high-school biology courses*. A few definitions of terms are in order here. By "most students" I mean perhaps the bottom 75%. I do not deny that the best-performing students are learning, and understanding, quite a lot from their biology courses. I define "useful" knowledge as knowledge that helps students do something other than pass tests—knowledge that they can use in out-of-school contexts. Although my convictions arise from the experiences described above, they are certainly not the only evidence of the truth of the above assertion. When Yager and Yager (1985) tested students' ability to select correct definitions of terms from the biological and physical sciences, they found evidence that seventh-graders did better than third-graders, but there was no improvement at all between seventh and eleventh grades, the time when most students take high-school biology! In the most recent studies of science achievement by the International Association for the Evaluation of Educational Achievement (1988), American high-school seniors were dead last among students in 13 countries ranked for assessed biological knowledge.

These results lead me to two questions about assessment. The first arises from the fact that our present assessment system has declared these students to be *successful* biology students. They almost all graduated in the top half of their class, and they passed their previous biology courses. Why don't our tests reveal to us how little they are really learning? Second, how can we do a better job? We need to assess biology learning in ways that both give us valid descriptions of students' knowledge and help us to improve the practice of biology teaching. The remainder of this paper addresses these questions.

CONCEALING STUDENTS' IGNORANCE

Why don't our present assessment procedures, including both teacher-made and standardized tests, do a better job of revealing to us how little students are learning? I do not believe that it is difficult to devise assessment procedures that will reveal students' lack of learning. Almost any question that requires students to write or speak entire sentences (and many questions that do not) will work. However, the demands of producing tests that maximize efficiency and reliability (along with the vested interest that many people have in *not* seeing how poorly our system is working for non-elite students) have led us to create an elaborate assessment system that could hardly have been better designed to conceal students' lack of knowledge. We could take from our present system a set of object lessons in how to draw attention away from the absence of significant student learning. In particular, our present assessment and reporting procedures incorporate the following practices, each of which helps to obscure the fact that students are hardly learning anything.

Don't Give Them Time to Forget

During one of our studies at the middle-school level, we had to administer posttests after students had completed each of three units. My colleague Ed Smith was discussing with one of the teachers when we should administer the posttest for a unit that his class would complete on a Thursday. "You'd better come on Friday," the teacher said. "There's no telling how much they will still remember by Monday." The most discouraging aspect of this story is that the teacher may well have been right. One characteristic of each of the studies cited above is that there was an appreciable delay between the time of instruction and the time of testing. Forgetting is probably partly responsible for the students' poor performance.

Does this mean that the tests weren't "fair"? Not at all, if the purpose of testing is to assess useful knowledge. When do we expect the occasions to arise when students will use their biological knowledge? Surely they won't all be in the first week, or the first month, after the relevant concepts have been taught.

The fact that deterioration of knowledge is a major problem is a sad commentary on our present biology curriculum, because it isn't a problem for everyone or for everything we teach. *I* haven't forgotten how to define "photosynthesis," even though my last biology course was before the last biology course of the students quoted above. The students studied reading and writing before they studied biology, but they haven't forgotten *that.*

The difference between the memories that we retain and the ones that deteriorate has a lot to do with the usefulness of the knowledge. We forget or jumble up useless facts, while we remember the concepts, principles, and skills that we use to interpret and operate in the world. It appears that the students quoted above are trying (and usually not quite succeeding) to remember facts, not intellectual tools that they are accustomed to using to interpret the living world around them. Students will appear to remember more of those memorized facts if they are tested right away, but we are being deceived by their performance if we conclude that they have gained useful knowledge.

Report Scores in Numerical Form

The study that produced the student test responses quoted above was designed to improve a pair of courses that were sometimes taught by me and sometimes taught by colleagues in the Natural Science Department at Michigan State University. I thought that if they just looked at what their students were saying, they would see the need for substantial revision in the courses' curriculum and instruction. I gave one professor a sample of 10 posttests and asked him to look at them. I met with him again several weeks later, and he gave the tests back to me. I asked if he had read them. "No, I didn't," he said. "I didn't know how you wanted them interpreted." This same professor looked regularly at the item analyses for his multiple-choice tests, however, and constructed hypotheses (some of which I believed to be substantially erroneous) about why students missed questions.

Numerical data reveal who is doing a little better, or who is doing a little worse, or how students are doing on the average. They do not tell us very much, however, about how students are thinking or what they know. Actual test items (or interview questions) and actual student responses, especially longer written responses that reveal student reasoning, tend to confront the reader with the qualitative reality of students' thinking. Many teachers and policy-makers would prefer to hide behind a veil of numbers that leave some distance between them and this reality.

Focus on Efficiency and Reliability at the Expense of Validity

Several years ago I attended a colloquium presentation by the person in charge of the design of the science section of the Stanford Achievement Test. He described an elaborate procedure by which the test was developed, moving from objectives to item pools, to item assessment, to the development of alternative forms of the test, and so forth. He used a single test item to illustrate this process, an item that ostensibly tested for student mastery of a "science process skill." It appeared to me that there were at least three ways to arrive at the correct response for this item, two of which did not involve use of the process skill at all. When I asked him how he knew what the item was really testing, he invoked the whole long test development process again. At no point during test development, however, did anyone ever ask a student how he or she arrived at an answer.

It seems to me that the above incident revealed some basic differences in our assumptions about the nature and purposes of science achievement tests. This test development process emphasized *efficiency* and *reliability*; it produced a machine-scorable test that produced consistent student scores. My concerns, however, focused on *validity*; I wanted to know what the scores meant.

The idea of using interviews to assess the validity of test items and scoring procedures is not original with me. Yarroch (1986) asked students how they arrived at their responses to items in the Michigan Educational Assessment Program science test. He found that students frequently were able to arrive at correct answers through incorrect reasoning. Less often, essentially correct reasoning led students to choose incorrect responses. Norris (in press) reviewed a series of studies using similar methods and generally arriving at similar conclusions: The reasoning that students actually use to arrive at responses to multiple-choice questions may be different from what the test developers assume it is. When test items are not revealed, it is difficult to assess how big a problem this is.

Apart from the issue of whether our current assessment procedures actually measure what they purport to measure is the question of whether what they purport to measure is useful knowledge. This question can be addressed at two levels. At one level, there is an issue of *face validity*: Does there seem to be a reasonable similarity between what we ask students to do on tests and what they might actually do with their knowledge in out-of-school contexts? At a deeper level, there is an issue of *construct validity*: Do the tests portray biological knowledge and learning in ways that are consistent with current scholarship in philosophy and psychology? On a more practical level, does the information obtained with current assessment procedures help us to develop appropriate policies or improve curriculum and instruction? In each respect, I believe that our current assessment procedures are lacking.

With regard to face validity, I would simply observe that in out-of-school contexts, we sometimes speak or write about the living world in sentences. Yet my students have told me that it is possible for a biology major to graduate from Michigan State University without ever having to write a sentence on a test. Once they graduate, they will be expected to use their knowledge differently. Not since my student days has anyone asked me to answer a multiple-choice question about biology.

With regard to construct validity, few people would actually claim that biological knowledge consists of a large number of independent and equally important bits. Yet when we give multiple-choice tests and treat the numbers of correct answers as interval data, this is precisely the assumption that is built into the technology of assessment. Many tests, both teacher-made and standardized, are, of course, accompanied by elaborate theoretical frameworks that describe biological knowledge in much more complicated terms. In this case, however, the medium often is the message. Students studying for a test, or teachers preparing their students, are likely to ignore the rhetoric and be guided in their preparations by the form of the test itself.

DEVELOPING BETTER ASSESSMENTS OF BIOLOGICAL KNOWLEDGE AND LEARNING

Criticizing current assessment practices is, of course, much easier than coming up with good alternatives. Nonetheless, a variety of alternative approaches to assessment have been developed. In this section, I will describe an approach that has been developed over the last 10 years by a research group at Michigan State University that includes my colleagues Ed Smith and Kathy Roth, several other professors and graduate students, and me. Other work in this area has been done by Rosalind Driver and colleagues at the University of Leeds (Driver and Erickson, 1983; Driver et al., 1985), James Stewart and colleagues at the University of Wisconsin (Stewart, 1983), and others.

The nature of our assessment procedures has been determined by the larger goals that they served. Our research and development program has had two essential goals. First, we have been interested in interpreting classroom instruction—in understanding how students and teachers act and talk in classrooms and why some instructional strategies work better than others. Second, we were engaged in developing improved teaching methods and materials. To accomplish these goals, we needed rich and detailed descriptions of students' knowledge and thinking that were consistent with our philosophical and psychological understanding. Thus, we were willing to sacrifice some efficiency and reliability for richness of description and construct validity.

Developing rich and psychologically sophisticated descriptions of students' knowledge is not an easy task. For example, although it is relatively easy to see that the students quoted above are deficient in formal biological knowledge, we wanted to go beyond that. We wanted to understand how they arrived at the responses that they gave. What *did* they know or believe, from whatever source, that led them to think as they did about the problems that we posed?

In our attempts to develop assessment procedures that answered the above question, we drew on scholarship from a number of sources. The first of these was the history and philosophy of science (e.g., Mayr, 1982; Toulmin, 1972), which provided important ideas about the nature of scientific knowledge, as well as about the metaphorical similarities between systems of human knowledge and biological systems. A second source was social constructivist psychological theory and work applying it to problems of education (e.g., Collins et al., in press; Rogoff and Lave, 1984; Vygotsky, 1962, 1978), which provided ideas about the relationship between individual knowledge and social interactions. Finally, we drew heavily on other work that, like ours, approached problems of science education from a constructivist or conceptual-change orientation (e.g., Driver et al., 1985; Posner et al., 1982; West and Pines, 1985).

Describing Students' Knowledge and Learning

In their present form, our assessment procedures consciously draw on biological metaphors (which we believe to be more appropriate than the computer metaphors that prevail in much cognitive scientific work) to describe human knowledge and learning. We think of human knowledge as consisting, like the living world, of many complex, interacting systems that can be characterized in terms of their structure, their functions, and the patterns of their development.

Structure

All human knowledge—even the knowledge of apparently confused students like those quoted at the beginning of this paper—is highly structured. Like some biological structures, the structures of human knowledge are complex and constantly changing. There is also an analogy in human knowledge to the hierarchically nested nature of biological structures. In particular, human knowledge has social, as well as individual, dimensions. Communities, including communities of scientists, work cooperatively to build knowledge structures that are far larger and more complex than any individual could ever master. The academic disciplines, including biology, are such socially constructed knowledge structures.

Describing students' knowledge involves recognizing the complex interrelationships among their ideas, including relationships that go outside disciplinary boundaries. Indeed, it appears that formal biological knowledge can truly be meaningful to beginning students only if they can relate it systematically to the many ideas about the living world (some correct, some incorrect) that they already had before they began the formal study of biology.

Although we find this characterization of the structures of students' biological knowledge to be metaphorically useful and consistent with current scholarship in philosophy and psychology, it suggests that the practical task of describing the knowledge structures of individual students is immensely difficult. Because these knowledge structures are so complex, diverse, and dynamic, in addition to being invisible, we have never seen a useful and practical approach to describing them. Our assessment procedures have therefore avoided attempts to develop complete descriptions of the structure of students' knowledge, relying instead on structural comparisons of the knowledge of different individuals (see the discussion of development below).

Functions

Biological facts, theories, and principles are not inert "content." They are more like intellectual tools or body organs, in that they have functions, as well as structures. In particular, biological knowledge helps us to describe, explain, make predictions about, and control the living world. Each of these functions is a social activity that involves the application of biological knowledge to living systems (see Anderson and Roth, in press).

Description, explanation, prediction, and control are not, however, functions exclusively of scientific knowledge. Even young children who have no exposure to formal science instruction engage in these activities by using their personal and cultural knowledge. Biological concepts and principles make it possible for us to engage in these activities with far more power and precision than would otherwise be possible. Thus, for us a critical test of students' understanding is their ability to use biological concepts and principles to describe, explain, predict, and control living systems. (Note that these functions of scientific knowledge are different from what other science educators sometimes refer to as science process skills or scientific thinking skills, in that the functions of scientific knowledge involve the use of existing knowledge, rather than the development of new knowledge.)

In our assessment procedures, we recognize the importance of the functions of scientific knowledge in the ways that we specify instructional objectives. The objectives always specify ways in which students should be able to use biological knowledge to describe, explain, predict, and control living systems.

Development

Biological knowledge is constantly changing, both in individuals and in communities. I am particularly attracted to the analogies that Toulmin (1972) draws between the development of scientific knowledge and processes of evolution or ecological succession. Toulmin speaks of an "intellectual ecology," in which individual concepts are seen as analogous to populations in an ecosystem. The intellectual ecology of an individual or a scientific community changes gradually, through processes involving both cooperation and competition among concepts. A new concept can "take root" and thrive only if a complex of other concepts on which it depends is already in place; thus, students go through stages of intellectual development analogous to stages in ecological succession. In the early stages of development, students depend primarily on concepts that are part of our common cultural knowledge base. At later stages, they are able to incorporate specialized scientific concepts and principles into their individual "intellectual ecologies" (see Posner et al., 1982).

In our work, we have tried to describe this process of intellectual development by drawing comparisons between students at different stages of development. Table 1 and Figures 1 and 2 show different ways that we have used to make those comparisons. Table 1 (from Anderson et al., 1987) is an example of our most common approach: a series of comparisons between common patterns in student thinking (Naive Conceptions) and the ways that we would like them to use scientific knowledge to think about the same issues (Goal Conceptions). Figure 1 (from Bishop and Anderson, in press) and Figure 2 (from Smith and Anderson, 1986) contrast naive conceptions and goal conceptions in diagram form.

We regard these conceptual analyses of students' thinking as the most important outcome of our assessment procedures. These procedures do, however, also produce numerical data for the purposes of making comparisons between different instructional treatments. In general, we report

TABLE 1 Respiration Issues and Conceptions

Issue	Goal Conception	Naive Conception
Implicit definition of respiration	Respiration is the process by which all cells obtain energy from food.	Respiration is breathing which occurs only in animals.
Nature of food	Food is matter that organisms can use as a source of energy.	Food is the stuff that organisms eat.
Function of food	Food supplies the energy that cells need for life processes.	Food keeps organisms alive.
Source of energy	The only source of energy for any organisms is the energy stored in food.	Organisms get energy from many different sources.
Energy transformation	Energy stored in food is released in a form that can be used by cells.	Food energy is used directly (no notion of energy transformation).
Matter transformation	Food is chemically combined with oxygen to create carbon dioxide and water, accompanied by the release of energy.	Food is digested and excreted. Oxygen is changed into carbon dioxide. These two processes are not related to one another.
Movements of reactants and products	Food and oxygen are supplied to all cells via the respiratory and circulatory systems. Carbon dioxide and water are removed from cells by these same systems.	Food goes to the stomach, gets digested and is excreted. Oxygen goes into the lungs and carbon dioxide comes out. (No notion of distribution to cells.)
Nature of energy	Energy changes from one form to another: light ➤ stored energy in food ➤ energy for life processes ➤ heat.	Energy is confused with matter, which contains energy, and gets used up (like fuel).

the percentage of students who demonstrated mastery of each goal conception (Anderson and Smith, 1986; Anderson et al., in press; Bishop and Anderson, in press).

Developing Tests and Analysis Procedures

The table and figures are products of a fairly long and complex development process that produces topic-specific tests and analysis procedures for each test. These procedures are described in detail in the cited references. Briefly, they involve the following steps:

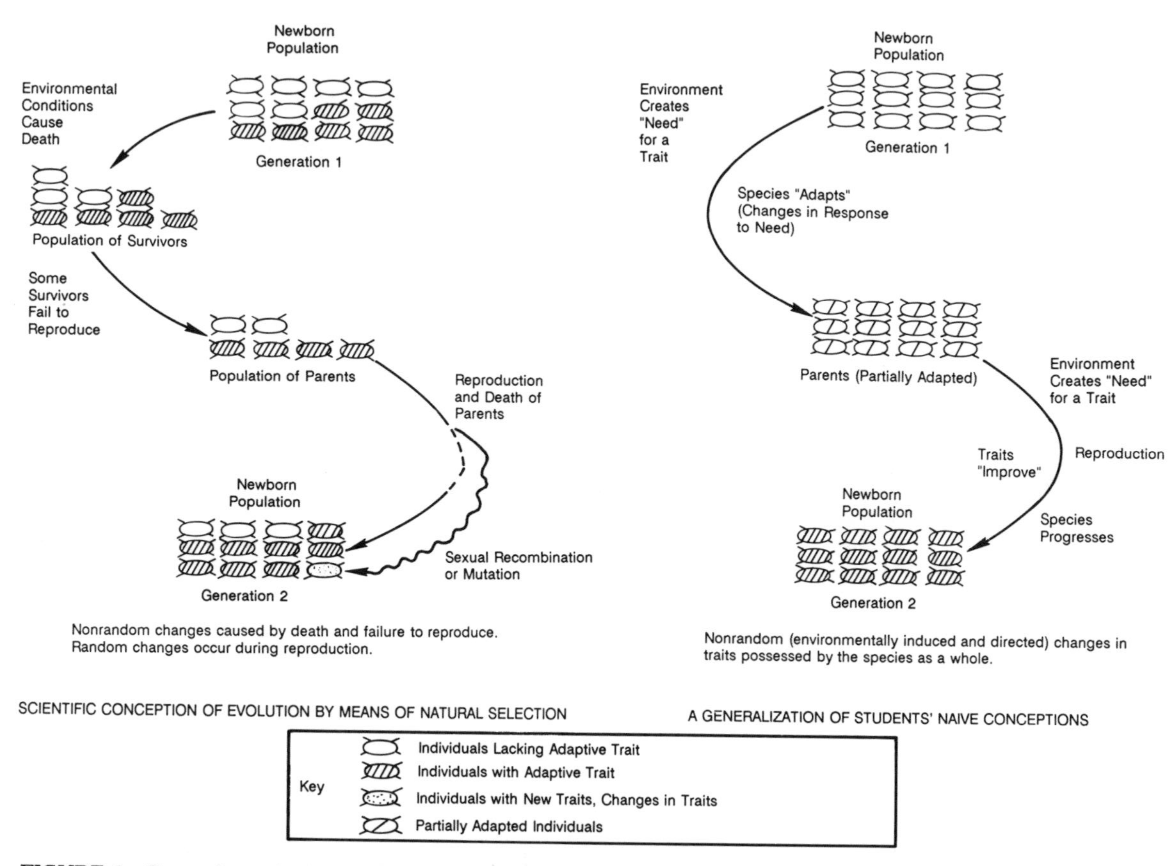

FIGURE 1 Comparison of scientific and naive understandings of the mechanism of evolution.

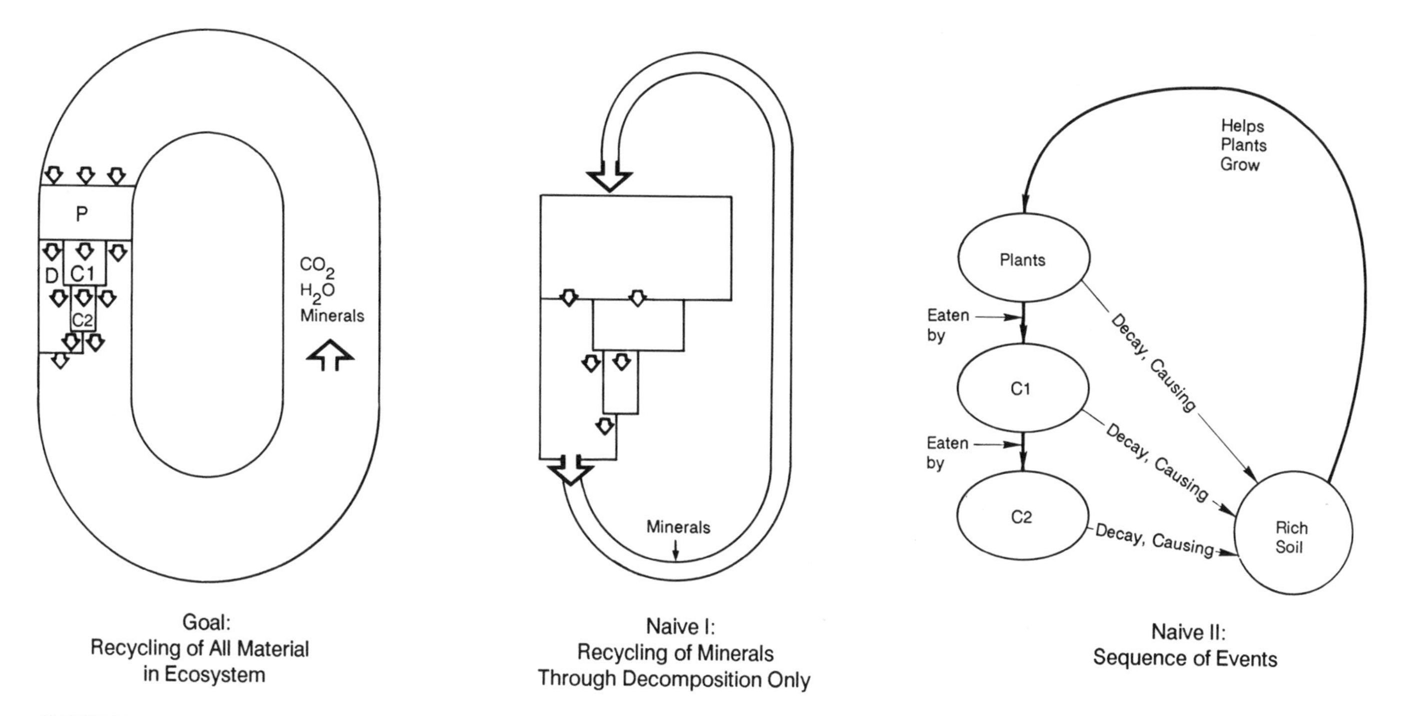

FIGURE 2 Conceptions of movement of matter in ecosystems.

• *Content analysis of the topic.* We begin with an analysis of the biological knowledge to be tested as it is presented in relevant texts and as we understand it ourselves. Although texts often present biological knowledge as consisting primarily of facts and vocabulary words, we are interested in how students could use biological concepts and principles as intellectual tools that help them to make sense of the living world. Thus, the outcome of our content analysis is a set of behavioral objectives specifying how students should be able to use knowledge of the topic to describe, explain, predict, and control the world around them.

• *Developing preliminary tests or clinical interviews.* The next stage in development is a preliminary exploration of student conceptions. In general, we begin this process by asking open-ended questions that require students to describe, explain, make predictions about, and control relevant living systems in their own terms. For example, early in the development of the photosynthesis test, we asked students to explain how a bean plant gets its food. The question about how cheetahs developed the ability to run fast (quoted at the beginning of this paper) is another example.

• *Developing questions that focus on critical issues.* The early open-ended questions lead us to hypotheses about critical differences between common student conceptions and canonical scientific conceptions. We test those hypotheses by developing and field-testing short-answer and multiple-choice questions that focus directly on the issue of interest. The question about energy for bean plants and humans is an example. The final test contains a mixture of open-response and forced-choice items.

• *Coding of student responses.* After tests are administered, we begin the analysis process by developing and using systems to code student responses for critical characteristics. Sometimes a single response is coded for more than one characteristic. For example, each response to the cheetah question above was coded for three characteristics: whether the origin of new traits is attributed to a random process, such as mutation and sexual recombination, or to the organism's response to the environment; whether diversity in the ancestral population is recognized and assigned a role in the evolutionary process; and what process is suggested to account for population change (see Bishop and Anderson, in press).

• *Analysis of coded responses.* Whether a student has mastered the goal conception for each issue (see Table 1) is assessed by calculating the weighted average of the evidence provided by all relevant questions. In general, a test contains several items relevant to each issue. Thus, the final outcome of the analysis is an assessment of the degree to which each student understands and is able to use each of the scientific goal conceptions.

Uses of the Assessment Procedures

As stated above, the assessment procedures described here were designed to be subordinate to other goals: the evaluation of classroom instruction and the development of improved teaching methods and materials. They have worked well for those purposes. The descriptions of students' knowledge have allowed us to understand their responses to classroom instruction in ways that would otherwise have been impossible (Anderson and Roth, in press; Smith and Anderson, 1984). Furthermore, they have helped us to develop teaching methods and materials, as well as approaches to teacher education, that are demonstrably more effective than those which currently prevail in school classrooms (see Anderson and Roth, in press; Roth and Anderson, 1987; Roth et al., 1988). In our more successful development efforts, we have raised the percentage of students showing mastery of goal conceptions from the 0-20% range (when teachers used commercial methods and materials) to the 50-80% range (when teachers used the methods and materials that we had developed).

CONCLUSION

Most current assessment procedures, including both teacher-made and standardized tests, rely on techniques that emphasize efficiency and reliability. There are good reasons for this: resources are limited, we want to be fair to all students, and we need accurate and reliable data for policy purposes.

However, in emphasizing efficiency and reliability, we have developed an array of assessment techniques that conceal students' lack of learning and that portray biological knowledge and learning in a woefully inadequate and distorted way. This is a tremendous price to pay for efficiency and reliability, inasmuch as the views of knowledge and learning built into our assessment techniques affect the thinking and behavior of teachers, students, and curriculum developers. In so doing, they contribute to the inadequacies and distortions of our present system of biological education.

I have briefly described an alternative approach to assessment (not the only one) that sacrifices some efficiency and reliability for construct validity and richness in description of students' knowledge. I believe that it is a good trade. By portraying student knowledge and learning in philosophically and psychologically more sophisticated ways, these assessment techniques focus attention on critical problems in biology teaching and help us to develop solutions to those problems. I believe that other assessment systems would benefit from a similar shift in emphasis.

REFERENCES

Anderson, C. W., and K. J. Roth. In press. Teaching for meaningful and self-regulated learning of science. In J. Brophy, Ed. Teaching for Meaningful Understanding and Self-Regulated Learning. Greenwich, Conn.: JAI Press.

Anderson, C. W., and E. L. Smith. 1986. Children's Conceptions of Light and Color: Understanding the Role of Unseen Rays. Research Series No. 166. East Lansing, Mich.: Institute for Research on Teaching, Michigan State University.

Anderson, C. W., K. J. Roth, R. E. Hollon, and T. D. Blakeslee. 1987. The Power Cell: Teacher's Guide to Respiration. Occasional Paper No. 115. East Lansing, Mich.: Institute for Research on Teaching, Michigan State University.

Anderson, C. W., T. H. Sheldon, and J. DuBay. In press. The effects of instruction on non-majors' conceptions of respiration and photosynthesis. J. Res. Sci. Teach. (also available as Institute for Research on Teaching Research Series No. 164)

Bishop, B. A., and C. W. Anderson. In press. Student conceptions of natural selection and its role in evolution. J. Res. Sci. Teach. (also available as Institute for Research on Teaching Research Series No. 165)

Collins, A., J. S. Brown, and S. E. Newman. In press. Cognitive apprenticeship: Teaching the craft of reading, writing, and mathematics. In L. B. Resnick, Ed. Learning and Knowledge: Issues and Agendas. Hillsdale, N.J.: Lawrence Erlbaum.

Driver, R., and G. Erickson. 1983. Theories-in-action: Some theoretical and empirical issues in the study of students' conceptual frameworks in science. Stud. Sci. Educ. 10:37-60.

Driver, R., E. Guesne, and A. Tiberghien. 1985. Children's Ideas in Science. Philadelphia: Open University Press.

International Association for the Evaluation of Educational Achievement. 1988. Science Achievement in Seventeen Countries: A Preliminary Report. Elmsford, N.Y.: Pergamon Press.

Mayr, E. 1982. The Growth of Biological Thought. Cambridge, Mass.: Belknap.

Norris, S. P. In press. Using studies of thinking processes to develop multiple-choice tests of empirical reasoning competence. In D. N. Perkins, J. Segal, and J. F. Voss, Eds. Informal Reasoning and Education. Hillsdale, N.J.: Lawrence Erlbaum.

Posner, G., K. Strike, P. Hewson, and W. Gertzog. 1982. Accommodation of a scientific conception: Toward a theory of conceptual change. Sci. Educ. 66:211-227.

Rogoff, B., and J. Lave, Eds. 1984. Everyday Cognition. Cambridge, Mass.: Harvard University Press.

Roth, K. J., and C. W. Anderson. 1987. The Power Plant: Teacher's Guide to Photosynthesis. Occasional Paper No. 112. East Lansing, Mich.: Institute for Research on Teaching, Michigan State University.

Roth, K. J., C. L. Rosaen, and P. E. Lanier. 1988. Mentor Teacher Project: Program Assessment Report. East Lansing, Mich.: Michigan State University.

Smith, E. L., and C. W. Anderson. 1984. Plants as producers: A case study of elementary school science teaching. J. Res. Sci. Teach. 21:685-695.

Smith, E. L., and C. W. Anderson. 1986. Alternative Student Conceptions of Matter Cycling in Ecosystems. Paper presented at the annual meeting of the National Association for Research in Science Teaching, San Francisco.

Stewart, J. 1983. Student problem solving in high school genetics. Sci. Educ. 67: 523-540.

Toulmin, S. 1972. Human Understanding. Princeton, N.J.: Princeton University Press.

Vygotsky, L. S. 1962. Thought and Language. Cambridge, Mass.: MIT Press.

Vygotsky, L. S. 1978. Mind in Society. Cambridge, Mass.: Harvard University Press.

West L. H., and A. L. Pines, Eds. 1985. Cognitive Structure and Conceptual Change. Orlando, Fla.: Academic Press.

Yager, R. E., and S. O. Yager. 1985. The effects of schooling upon understanding of selected science terms. J. Res. Sci. Teach. 22:359-364.

Yarroch, W. L. 1986. Content Validity of the 1985 Michigan Department of Education Pilot Science Examination. Paper presented at the annual meeting of the National Association for Research in Science Teaching, San Francisco.

9
The Advanced-Placement Biology Examination: Its Rationale, Development, Structure, and Results

WALTER B. MACDONALD

The advanced-placement (AP) biology course sponsored by the College Entrance Examination Board (College Board) is a national program that provides an opportunity for high-school students to pursue and receive credit for college-level biology coursework. The program is intended to replace biology courses that would normally be taken at the freshman or sophomore level in college and is based on the premise that college-level material can be taught successfully to able and well-prepared high-school students. The AP biology course is open to any high school that elects to participate; similarly, the AP biology examination is open to any student who wishes to take it. The AP examinations are administered once a year, in May, under standardized conditions at participating schools in the United States and many other countries. Most students take AP examinations in their own schools; others take them in multischool centers.

DEVELOPMENT OF THE AP BIOLOGY COURSE AND EXAMINATION

The policies of the AP biology course and examination, like those of the AP courses and examinations in other subjects, are determined

Walter B. MacDonald received his Ph.D. in ecology in 1983 from Rutgers University. Since joining the Educational Testing Service (ETS) in 1984, he has served as coordinator of the College Board's biology achievement test, as science coordinator of the College Board's Educational Equality Project, and as the College Board's member of ETS's Test Development Document Creation (TD/DC) project. Currently, Dr. MacDonald is director of test development for the National Assessment of Educational Progress.

by representatives of College Board member institutions and agencies throughout the country, including public and private high schools, colleges, and universities. The preparation of the course is an ongoing process, and the design of each examination typically begins nearly 2 years before the actual administration. Operational aspects of the examination—including the development of materials, scoring, and grading—are managed by the Educational Testing Service (ETS).

The AP Biology Development Committee, appointed by the College Board, is the "heart and mind" that prepares both the course description and the examination itself. The committee is made up of college professors and high-school AP teachers; these individuals are familiar with the academic standards to which college freshmen or sophomores are held. The committee is the authority for subject-matter decisions that arise in the test construction process. Committee members bring to their tasks knowledge of biology curricula and of laboratory methods; they are cognizant of the abilities and understandings that are critical to mastery of biology and how students might be asked to demonstrate these abilities and understandings.

COURSE DESCRIPTION

The AP biology course is taught by high-school biology teachers with guidance from *Advanced Placement Course Description—Biology*, a College Board publication prepared by the Development Committee. The course description provides broad guidelines for the content and skills to be included in the course and offers a recommended set of quantitative laboratory exercises. In addition, the publication contains information about the examination, sample questions illustrative of those included in the examination, a list of recommended textbooks, and other materials and resources helpful in preparing and teaching a college-level biology course to high-school students. AP biology teachers also receive assistance in developing and teaching their courses from other publications and from workshops and special conferences.

Biology is a dynamic science; over the last few years, many new areas of inquiry have come to the forefront, while others previously emphasized in the discipline have receded. Sensing the need to reassess the content of biology instruction at the college level, the Development Committee surveyed the introductory biology courses at more than 80 colleges and universities across the country. Its primary goal was to obtain current information on what is taught so that the AP biology course syllabus could be revised to reflect collegiate course offerings more closely.

On the average, respondents participating in the survey were able to categorize about 99% of what they taught into the 10 major biological categories presented in a questionnaire. The average percentage of course time devoted to each of these categories and the average emphasis placed on 72 other subcategories and topics provided the information needed to update the AP biology curriculum.

The survey of colleges and universities also affirmed the view that laboratories still are a central part of biology instruction and that certain topics are covered in these laboratories with great frequency. To maintain parity between AP biology and college-level courses, the Development Committee has included sample laboratory experiments in the description booklet to augment any other laboratory experiments already taught by AP teachers. Most of the experiments in the booklet are patterned after those often included in colleges. The current AP biology topics, the approximate percentages of emphasis, and the topics of 12 laboratory experiments are outlined in Figure 1.

STRUCTURE OF THE EXAMINATION

The AP biology examination is 3 hours long and is designed to measure a student's knowledge and understanding of college-level biology. The examination consists of a 90-minute multiple-choice section with 120 questions that examines the learning of representative facts and concepts drawn from across the entire curriculum and a 90-minute, free-response section consisting of four mandatory questions that address broader topics. The number of multiple-choice questions taken from each major topic of biology reflects the weighting of that topic as designated in the course syllabus. In the free-response portion of the examination, one essay question focuses on molecules and cells, one on genetics and evolution, and two on organisms and populations. Any of these four questions may require the student to analyze and interpret data or information drawn from laboratory experiences, as well as lecture material; to design experiments; and to demonstrate the ability to synthesize material from several sources into a cogent and coherent essay. To allow students to show their mastery of laboratory science skills and knowledge, some questions in the multiple-choice section and one of the four essay questions may reflect the laboratory work and the objective associated with the AP biology laboratory exercises.

The multiple-choice section of the examination counts for 60% and the free-response section 40% of the student's grade. In order to provide maximal information about differences in students' achievements in biology, the examinations are intended to have average scores of about 50-60% of the highest possible score for the multiple-choice and free-response sections.

Using questions written by college faculty and AP teachers, the examination is assembled by ETS consultants to both content and statistical specifications. Each examination contains both new questions and a set of questions that have been included on previous examinations. The set of

	TOPICS	PERCENTAGE GOALS	LAB TOPICS
I.	Molecules and Cells	25%	
	A. Biological chemistry	7%	
	1. Review of atoms, molecules, bonding, pH, water		
	2. Carbon, functional groups		1. Diffusion and Osmosis
	3. Carbohydrates, lipids, proteins, nucleic acids		
	4. Chemical reactions, free-energy changes, equilibrium		
	5. Enzymes: coenzymes, cofactors, rates of activity, regulation		2. Enzyme Catalysis
	B. Cells	10%	
	1. Prokaryotic and eukaryotic cells		
	2. Plant and animal cells		
	3. Structure and function of cell membranes		
	4. Structure and function of organelles, subcellular, components of motility, cytoskeleton		
	5. Cell cycle: Mitosis, cytokinesis		3. Mitosis and Meiosis
	C. Energy transformations	8%	
	1. ATP, energy transfer, coupled reactions, chemiosmosis		
	2. C_3 and C_4 photosynthesis		4. Photosynthesis
	3. Glycolysis, fermentation, aerobic respiration		5. Cell Respiration
II.	Genetics and Evolution	25%	
	A. Molecular genetics	9%	
	1. DNA: structure and replication		
	2. Eukaryotic chromosomal structure, nucleosome, transposable elements		
	3. RNA: transcription, mRNA editing, translation		
	4. Regulation of gene expressions		
	5. Mutations		
	6. Recombinant DNA, DNA cloning, hybridization, DNA sequencing		6. Molecular Biology
	7. DNA and RNA viruses		
	B. Heredity	8%	
	1. Meiosis		
	2. Mendel's laws, probability		
	3. Inheritance patterns: chromosomes, genes, alleles, interactions		7. Genetics of *Drosophila*
	4. Human genetic defects		
	C. Evolution	8%	
	1. Origin of life		
	2. Evidence for evolution		
	3. Natural selection		
	4. Hardy-Weinberg principle, factors influencing allelic frequencies		8. Population Genetics and Evolution
	5. Speciation: isolating mechanisms, allopatry, sympatry, adaptive radiation		
	6. Patterns of evolution, gradualism, punctuated equilibrium		
III.	Organisms and Populations	50%	
	A. Principles of taxonomy and systematics, five-kingdom system	1%	
	B. Survey of Monera, Protista, and Fungi	2%	
	C. Plants	15%	
	1. Diversity; classification, phylogeny, adaptations to land; alternation of generations in moss, fern, pine, and flowering plants		
	2. Structure and physiology of vascular plants		9. Transpiration
	3. Seed formation, germination, growth in seed plants		
	4. Hormonal regulation of plant growth		
	5. Plant response to stimuli: tropisms, photo-periodicity		
	D. Animals	23%	
	1. Diversity; classification, phylogeny, survey of acoelomate, pseudocoelomate, protostome, and deuterostome phyla	7%	
	2. Structure and function of tissues, organs, and systems (emphasis on vertebrates), homeostasis, immune response	10%	10. Physiology of the Circulatory System
	3. Gametogenesis, fertilization, embryogeny, development	5%	
	4. Behavior	1%	11. Habitat Selection
	E. Ecology	9%	
	1. Population dynamics, biotic potential, limiting factors		
	2. Ecosystem and community dynamics: energy flow, productivity, species interactions,succession, biomes		12. Dissolved Oxygen and Primary Productivity
	3. Biogeochemical cycles		

FIGURE 1 College Board's advanced-placement biology course and laboratory syllabus. From College Entrance Examination Board, 1988.

previously used questions, called the equating set, is a "mini-test" assembled to both the content and statistical specifications for AP biology. The use of an equating set enables a new test to be equated to past tests. The analysis of student performance on an equating set allows statisticians to predict how previous AP students would have performed on a new examination or how the current AP students would have performed on past examinations. All the new questions used on a test are pretested on college students across the nation. The use of both pretested new questions and the equating set provides the statistical data to maintain an examination that is appropriate for college-level biology while keeping the level of examination difficulty relatively constant from year to year.

While the multiple-choice section of the examination is machine-scored, the free-response section is hand-scored by over 100 readers chosen from among college and high-school biologists nationwide who are actively involved in introductory college-level biology courses or an AP equivalent. The training of readers ensures uniformity of grading and strict adherence to carefully developed standards. All essays are graded on a 10-point scale.

The free-response score and the multiple-choice score are weighted and summed to produce a composite score with a 150-point maximum. Students are then assigned a grade of 5 to 1 based on a detailed analysis of the total scores for all students, on equating data from previously tested AP biology students, and on correlation checks to ensure test reliability. A score of 5 indicates that a student is extremely well qualified to pursue upper-level college biology courses, whereas a grade of 3 indicates average preparation.

RESULTS

Over the last 10 years, the number of students taking the AP biology examination has increased by about 11% per year. In 1988, about 31,000 students took the examination, compared with about 11,000 in 1978. In 1988, over 3,000 schools offered an AP biology course. That year, scores were sent to more than 1,000 colleges across the country. The results by sex, grade, type of school, and ethnicity are displayed in Table 1.

Over the years, the AP biology examination has maintained a relatively constant level of difficulty. The data from the equating set tend to indicate that recent populations of AP biology students are slightly less able than past populations. The mean score has steadily declined from 3.35 in 1981 to 3.05 in 1988. The reported grades for AP students since 1981 are displayed in Figure 2. For 1988, 25.2% of the students scored 3, 23.5% scored 2, 23.4% scored 4, 15.4% scored 5, and 12.5% scored 1; thus, 64% scored 3 or higher. Over the last 8 years, the percentage of students who received a score of 4 has remained relatively constant, the percentage at 5 has slightly

TABLE 1 1988 National Summary Data for Biology[a]

		All Students	Female Students	Male Students	11th Grade	12th Grade
Total	N	30,612	15,653	14,959	8,613	19,320
	Mean	3.05	2.87	3.23	3.14	3.01
Black Students	N	1,165	770	395	294	819
	Mean	2.17	2.07	2.96	2.24	2.12
White Students	N	22,099	11,405	10,694	6,036	14,408
	Mean	3.04	2.87	3.09	3.09	3.02
Asian Students	N	3,875	1,746	2,129	1,294	2,104
	Mean	3.39	3.23	3.51	3.52	3.33
Hispanic Students	N	872	449	423	241	559
	Mean	2.56	2.37	2.76	2.62	2.51

[a]Data from College Entrance Examination Board, 1987.

decreased, the percentage at 3 has greatly decreased, and the percentages at 2 and 1 have increased.

The AP biology program has experienced tremendous growth over the last few years. There now are more students earning scores of 5, 4, and 3 than in the past. Unfortunately, many more students are earning scores of 2 or 1. This increase in the percentages of students at 2 and 1 may be due to the addition of many new schools with novice AP biology teachers. While it is rewarding to teach a college-level course in high school, it is not easy. Often it takes time for the novice AP biology teacher to develop the skills, level of preparedness, and enthusiasm typical of the veteran AP biology teacher.

What topics do students who score a 2 or 1 not fully understand? A review of the multiple-choice questions on recent examinations shows that many of these students fail to comprehend such basic topics as osmosis, plant-animal cell differences, function of cell organelles, differences between photosynthesis and respiration, DNA replication, RNA transcription, meiosis, inheritance patterns, natural selection, blood circulation, digestion, antigen-antibody relationships, and phylogenetic relationships. Most students who score a 3 show average understanding of these topics, whereas students who score 4 or 5 exhibit excellent understanding of these topics and many others.

In 1987 and 1988, many of the students who scored a 2 or 1 could not score more than 1 on essay questions that asked them to:

> Describe the production and processing of a protein that will be exported from a eukaryotic cell. Begin with the separation of the messenger RNA from the DNA template and end with the release of the protein at the plasma membrane [Educational Testing Service, 1987].
>
> or
>
> Discuss Mendel's laws of segregation and independent assortment. Explain how the events of meiosis I account for the observations that led Mendel to formulate these laws [Educational Testing Service, 1988].
>
> or
>
> Discuss the processes of cleavage, gastrulation, and neurulation in the frog embryo; tell what each process accomplishes. Describe an experiment that illustrates the importance of induction in development [Educational Testing Service, 1988].

Most students who scored a 3 could adequately answer these essay questions, whereas the students who scored a 5 or 4 were more likely to write more elegant and complete answers.

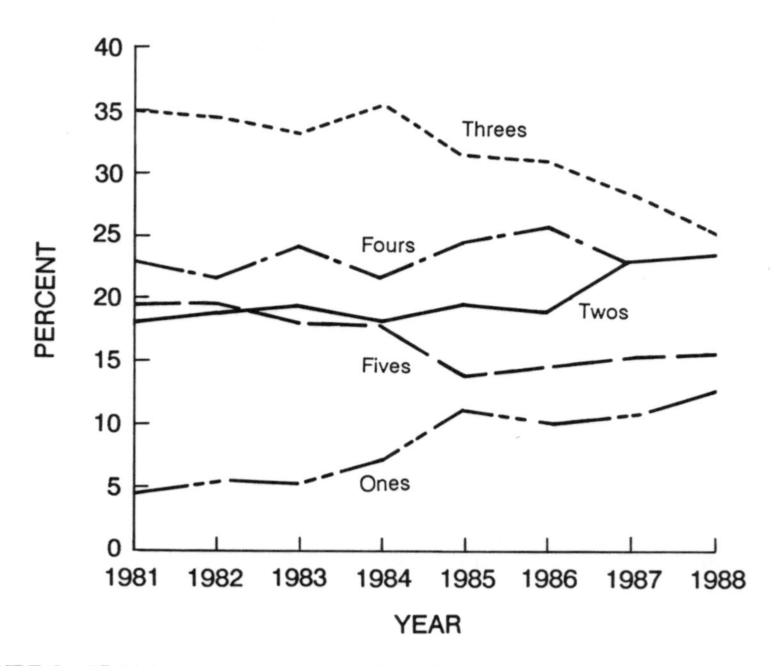

FIGURE 2 AP biology reported scores, 1981-1988.

SUMMARY

Generally, the state of the AP biology program and the status of AP biology students are very good. The majority of students are receiving a sound college-level course while attending high school. Validity and longitudinal studies have indicated that AP students perform as well as and often better than students taking the college course. Another important finding is that AP students tend to demonstrate higher achievement in college than their non-AP counterparts (Casserly, 1986). While one might expect AP candidates to show higher achievement in college than non-AP students because AP candidates are, in general, more able students, many AP candidates are placed in higher-level, more demanding courses when they reach college. Studies also have shown that 90% of the students who were placed ahead felt well prepared for the advanced sequence of college-level courses in which they then enrolled (Casserly, 1968). Other longitudinal studies have shown good correlation between scores on the AP biology examination and subsequent grades in introductory and upper-level biology courses in college (Willingham and Margaret, 1986).

An area of current and future concern is the increasing percentages of students who score 1 or 2 on the examination. It is hoped that the teacher preparation required to teach a college-level course in high school will catch up with the swelling population of AP students and precipitate a decrease in the percentages of students receiving scores of 1 and 2, while increasing the percentages of students receiving scores of 3, 4, and 5.

While participation by minority-group students in AP courses has increased over the last few years, a greater effort should be made to increase the participation of black and Hispanic students in AP biology courses. AP courses are a rewarding challenge that should be made available to all able students.

REFERENCES

Casserly, P. L. 1968. What college students say about advanced placement. College Board Rev. 69: Fall.

Casserly, P. L. 1986. Advanced Placement Revisited. College Board Report 86-2. New York: College Entrance Examination Board.

College Entrance Examination Board. 1987. Advanced Placement Program—National Summary Reports. New York: College Entrance Examination Board.

College Entrance Examination Board. 1988. Advanced Placement Course Description—Biology May 1989. New York: College Entrance Examination Board.

Educational Testing Service. 1987. The 1987 AP Biology Examination. Princeton, N.J.: Educational Testing Service.

Educational Testing Service. 1988. The 1988 AP Biology Examination. Princeton, N.J.: Educational Testing Service.

Willingham, W., and M. Margaret. 1986. Four Years Later: A Longitudinal Study of Advanced Placement Students in College. College Board Report 86-2. New York: College Entrance Examination Board.

10

The Development of Interest in Science

JON D. MILLER

There is broad agreement in the United States that scientific literacy is a good thing, that we don't have enough of it, and that it is especially important for our young people to have a lot more of it. There is also broad agreement that schools are the place where young people should get their scientific literacy and that formal institutions of education are failing to produce a minimal level of scientific literacy in an acceptable proportion of our young people. Most Americans, to borrow from the Declaration of Independence, find these truths to be self-evident. As educators and scientists, however, we cannot accept truths as self-evident, but must seek to understand better the roots of scientific literacy and the social, economic, and political consequences of scientific illiteracy.

Having spent the last several years studying data about young people's knowledge of and attitude toward science and having talked with a large

---

Jon D. Miller is professor of political science and director of the Public Opinion Laboratory, Northern Illinois University. He received an A.B. from Ohio University, an M.A. from the University of Chicago, and a Ph.D. in political science from Northwestern University. Dr. Miller directs research focused on the development of political and social attitudes in young adults and adults and on political behavior in democratic systems. He directs a major longitudinal study of the development of interest and competence in science and mathematics among middle-school and high-school students in the United States.

The Longitudinal Study of American Youth, including the work reported in this paper, is supported by National Science Foundation grant MDR-8550085. All the analyses, opinions, and conclusions offered are those of the author and do not necessarily reflect the views of the National Science Foundation or its staff.

number of students and teachers, I am convinced that functional scientific literacy requires some level of formal science and mathematics instruction. Informal learning programs like museums and television shows can augment formal instruction and stimulate interest in it, but these informal experiences cannot effectively replace or substitute for formal science instruction. Further, it is clear to me that functional scientific literacy requires the ability to read about science and technology to be able to sustain literacy in the decades after the end of formal instruction.

The basic problem is that formal instruction in science and mathematics has become voluntary in most American high schools and that attitudes have developed that discourage the vast majority of young Americans from attempting formal coursework in chemistry, physics, and mathematics beyond first-year algebra. Only 15% of last year's American high-school graduates had completed a physics course during their high-school experience, and only 30% had taken a chemistry course. Forty-five percent had avoided any contact with algebra throughout their 4 years of high school. And these figures apply only to students who graduated, excluding the sizable proportion that dropped out before graduation. Further, the data indicate that young women avoid science and mathematics at almost double the rate of young men. The problems are serious.

As scientists and educators, we must ask why so many young Americans decide not to study science and mathematics during their high-school years, and it is this question that has driven most of my recent work in this area. It is critically important that we come to understand the reasons for this pattern of science and mathematics avoidance. The British government recently addressed this problem by mandating that all British students take science and mathematics every year that they are in school and that they be tested through a national testing program to measure results. Compulsion is one solution, but with 16,000 independent school boards in the United States, compulsion is not an alternative available to us, regardless of its merits. If we are to do a good job with science and mathematics education in the United States, we must first understand the root sources of the attitudes of young Americans toward science and mathematics and seek to address those issues effectively.

The Longitudinal Study of American Youth (LSAY) is one effort to understand better the process of socialization and development of attitudes toward science and technology and citizenship. The LSAY builds on a previous cross-sectional study by Miller et al. (1980) and on the relevant literature. The LSAY will follow a national sample of seventh-graders and a parallel sample of tenth-graders for the next 4 years, collecting data from the students, their parents, their teachers, and related school staff. The base-year student data collection for the LSAY was completed during the 1987-1988 school year.

MEASUREMENT OF INTEREST IN SCIENCE

One approach to the problem of student avoidance of science is to examine the general attitude of students toward science. Apart from courses or specific encounters with science, most students have a general attitude or disposition toward science. The data collected by the LSAY provide an opportunity to construct a unidimensional measure of attitude toward science and to seek to understand the factors that contribute to fostering that attitude.

The base-year LSAY data collection included a series of items designed to tap each student's general attitude toward science. The full set of items was examined by both factor analysis and reliability tests, and the following five agree-disagree items were identified as a unidimensional measure of attitude toward science:

I enjoy science.
I am good at science.
I usually understand what we are doing in science.
Science is useful in everyday problems.
I will use science in many ways as an adult.

Each student was asked to strongly agree, agree, disagree, or strongly disagree with each of these items. An analysis of the marginal distribution of these items found a generally positive attitude toward science (see Table 1). A solid majority of high-school sophomores agreed that they liked science and felt that they understood it, but just over one-third thought that science would be helpful in their adult activities. The combination of these five items constitutes an index of general attitude toward science. The index is simply the number of agreements (strong or regular) with this set of items. The distribution of the students across the range of 0-6 was relatively even, with about one-third of the students scoring 0 or 1 on the index, one-third scoring 2 or 3, and one-third scoring 4 or 5. The mean score was 2.4, and the median score was 3.

It is likely that students scoring high on this index will be more likely to take advanced science courses and to engage in more informal science learning activities than students who score low. While we will have to await the second and third cycles of the LSAY for the individual change data to test that hypothesis, it is possible to use the base-year data to understand better the influence of home, school, and each student's life goals on his or her general attitude toward science.

SOME FACTORS ASSOCIATED WITH ATTITUDE TOWARD SCIENCE

While the aggregate distribution of student attitudes toward science is interesting, it is important to know more about which students hold

TABLE 1 Distribution of 2,829 Tenth-Grade Students on Five Attitude Items, 1987

	Strongly Agree (%)	Agree (%)	Not Sure (%)	Disagree (%)	Strongly Disagree (%)
I enjoy science.	17	40	14	19	9
I am good at science.	13	41	20	18	7
I usually understand what we are doing in science.	12	47	17	15	6
Science is useful in everyday problems.	8	29	35	20	6
I will use science in many ways as an adult.	11	24	39	16	9

more positive attitudes and which students hold more negative attitudes. We would expect that students who hold positive attitudes toward science would be most likely to enroll in advanced science courses and to carry more information from their courses into adulthood. We cannot test those hypotheses until we have obtained additional cycles of measures from the LSAY sample, but we can examine some of the characteristics associated with holding a positive attitude toward science among high-school sophomores.

The most proximate source of influence on a student's general attitude toward science might be expected to be the science course in which he or she is enrolled. While students—like all citizens—experience a wide array of technologies in their daily lives, it is primarily through formal science classes that students encounter science. Of course, some students may also experience science in science museums or on television or in books, but for most students, those experiences are far less frequent than class experiences.

As a starting point, it is useful to recall that most high schools require only 2 years of science and that not all tenth-grade students are enrolled in a science course. Most high schools offer students some choice in science courses, and some students elect to take a general science course while others move directly into biology. An examination of the course-taking patterns of the LSAY sophomore cohort found that 84% were enrolled in a science course and that 59% had enrolled in a biological science course, which is almost universally taught as a laboratory course at the high-school level. An additional 13% were enrolled in chemistry, and 1% were in a physics course; both are usually taught as laboratory sciences. Seven percent of sophomores were enrolled in a physical-science course, and 4%

in a general science course, neither of which normally involves extensive student laboratory work.

An examination of course-taking patterns by student sex, student educational expectations, and parental educational achievement indicated that sophomores with clear intentions to complete at least a baccalaureate were significantly more likely to be enrolled in chemistry than students without college aspirations. Sophomores not planning to go to college were more likely to be enrolled in general science, physical science, or no science at all. There appears to be no significant sex difference when parent education and student educational aspiration are held constant.

Beyond enrollment, it is important to know what each student thinks about the science courses to which he or she is exposed. In the LSAY, each student was asked to list each course that he or she was taking in the fall semester and to rate each course on eight dimensions (interest in subject of course, clarity of teacher, clarity of textbook, difficulty of course, whether course challenged student to think, likely utility of course to student's expected occupation, use of computers, and number of hours of homework each week). An examination of the data from the tenth-grade cohort indicated a substantial level of variance in these measures, suggesting that students were able and willing to differentiate among the different facets of each course and among the courses they were taking.

A factor analysis indicated that two of the dimensions captured a general attitude toward the course. One dimension concerned the student's interest in the subject matter of the course. A second dimension concerned each student's perception of the likely utility of the course in his or her career. As to interest in subject matter, students were asked to grade their level of interest in letter-grade terms. An A denoted a high level of interest in the subject matter, a C denoted an average level of interest, and an F denoted little or no interest. As to perceived utility, the same grade-card scoring was employed, with A meaning that the student thinks the course will be very useful in his or her career and F meaning that the course would be of no use. All five letter grades were available for use. For this analysis, I have converted these responses into traditional grade-point averages (GPAs), assigning 4 points for an A, 3 for a B, and 0 for an F. The index of attitude toward science course is the mean grade given by each student on the interest and utility dimensions.

Using this index, the tenth-grade cohort appears to hold generally positive attitudes toward their science courses. Using the same parent education, student aspiration, and sex context used to examine enrollment, the LSAY data indicate that high-school sophomores assign a B- to their science courses (see Table 2). In contrast to the 2.7 GPA assigned to science courses, the same sophomore cohort assigned GPAs of 3.1 to English, 2.9 to mathematics, and 2.5 to social studies. Science courses, it would appear,

TABLE 2 Evaluation of Science Courses by a National Sample of Public-High-School Sophomores, 1987

Parents' Education	Student's Expected Education	Student's Sex	Mean Score for Biology	Mean Score for Chemistry	Mean Score for Other Science
High school or less	Less than college	Male	2.2 (226)	2.3 (12)	2.6 (87)
		Female	2.3 (206)	2.5 (21)	2.3 (82)
	College degree	Male	2.6 (128)	2.8 (30)	2.6 (34)
		Female	2.6 (142)	2.9 (35)	2.6 (27)
	Graduate degree	Male	3.2 (76)	2.8 (32)	2.4 (10)
		Female	3.1 (134)	3.1 (44)	2.9 (16)
College degree	Less than college	Male	1.8 (28)	[a] (3)	[a] (8)
		Female	2.4 (43)	[a] (2)	2.5 (19)
	College degree	Male	2.7 (87)	3.1 (30)	2.9 (24)
		Female	2.3 (86)	3.0 (35)	2.3 (20)
	Graduate degree	Male	3.1 (94)	3.2 (47)	2.9 (16)
		Female	3.0 (98)	2.8 (45)	3.3 (18)
All public-high-school sophomores			2.6 (1,348)	2.9 (336)	2.6 (361)

[a] Too few cases available to calculate a reliable mean.

are viewed by sophomores more positively than social studies and less positively than English or mathematics. All four distributions, however, approximate normality, suggesting that some students hold very positive and very negative attitudes toward all four course areas.

Looking at the distribution of student attitudes toward science courses within the same parent education, student aspiration, student sex framework used to examine science-course enrollments, it appears that students expecting to complete a graduate degree hold the most positive attitudes toward science and that there are no systematic sex differences. While these multivariate tabulations are helpful in providing general impressions of the influence of each of these variables on the distribution of student attitudes toward science courses or toward science generally, we would like to know both the absolute and relative influence of each of these (and perhaps other) variables on student attitude toward science. It is possible to obtain a more precise measurement of the relationship of these and other background variables to student attitudes through the construction of a path model.

A MULTIVARIATE MODEL

The primary variable of interest to us is each student's general attitude toward science. A five-item index of student attitude toward science was introduced above, and the multivariate model will seek to understand the influence of several independent variables on the distribution of this attitude. The trichotomous distribution of the index described earlier will be used in this model.

In the preceding section, we identified student attitude toward science courses as the most proximate independent variable and looked at the distribution of student attitudes toward science classes. For this model, a single index of attitude toward science courses has been constructed, using the mean value of the attitude toward the science class in which the student is enrolled. In a very few cases, students were enrolled in more than one science class simultaneously, and in those cases the mean rating of the more advanced science class was used in the index. Approximately 400 sophomores were not enrolled in any science class, and they have been dropped from this analysis. Among the approximately 2,800 sophomores enrolled in a science course, 43% gave the course a C or lower, 29% a B, and 29% an A.

In the preceding tables, we have examined the distribution of student attitudes by the level of parental education, the level of education each student expects to complete, and the sex of the student. All three of these variables will be retained in the construction of a path model. The level of parent education will be dichotomized into less than a baccalaureate and the completion of a baccalaureate or more; 32% of the parents included in this analysis held a baccalaureate or more. The level of education expected by each student will be dichotomized into less than a baccalaureate and a baccalaureate or more; 66% of the sophomores included in this analysis expect to earn a baccalaureate or more. Fifty-two percent of the students in this analysis are female.

To explore more fully the impact of family practices and values on each student's attitude toward science, two additional variables will be added to the model. The first variable seeks to measure the degree to which parents encourage—or push—science. Each LSAY student was asked to mark a series of statements about his or her parents' attitudes and behaviors. A factor analysis of this battery of items identified five items that characterize parent science push. The index of parent science push is the number of student agreements with the following statements:

My parents want me to learn about computers.
My parents have always encouraged me to work hard on science.
My parents buy me mathematics and science games and books.
My parents expect me to do well in science.

My parents think that science is a very important subject.

For this analysis, the index of parent science push was dichotomized into parents who were reported by their student to do three or more of the five activities and parents who were reported to do fewer. Forty percent of the parents in the study were reported to do three or more of the science push activities.

The second family variable concerns the religious values of the parents. One parent from each LSAY family was interviewed by telephone in the spring of 1988, during the second semester of each student's sophomore year. A small set of religious-value questions were asked and subsequently used to create a typology of religious values. Parents who agreed with both the following statements were classified as religious conservatives:

There is a personal God who hears the prayers of individual men and women.
The Bible is the actual word of God and is to be taken literally, word for word.

Parents who disagreed with one or both of these statements were classified as religious moderates or liberals. For this analysis, parents were dichotomized into religious conservatives and others. Fifty-seven percent of LSAY parents were classified as religious conservatives. If there is a perceived conflict between science and religious values, we would expect to find it occurring most often among religious conservatives.

The inclusion of student course attitude, student educational aspiration, parent science push, parent religious attitudes, parent education, and student sex in a single model allows the exploration of the relative influence each of these measures—in the context of the relative impact of all the other variables—on each student's general attitude toward science. A path model (Goodman, 1978; Fienberg, 1980) is a convenient method of looking at these relationships and displaying the results in a relatively comprehensible format.

The path model to predict students' general attitude toward science indicates an interesting network of direct and indirect influences (see Figure 1). The variables are placed in an approximate temporal order. The current student attitude toward science is the object of our concern and the predictive object of the model. Student course experience is the most proximate independent variable and is placed closest to the dependent variable. Student educational aspirations may be thought of as having been formed before the immediate experience of courses and as being somewhat longer-standing in nature. This variable, therefore, is placed to the left of student course attitude, but to the right of the other variables. Similarly, parent science push may be viewed as of longer standing and likely before the formation of student educational aspirations. Student

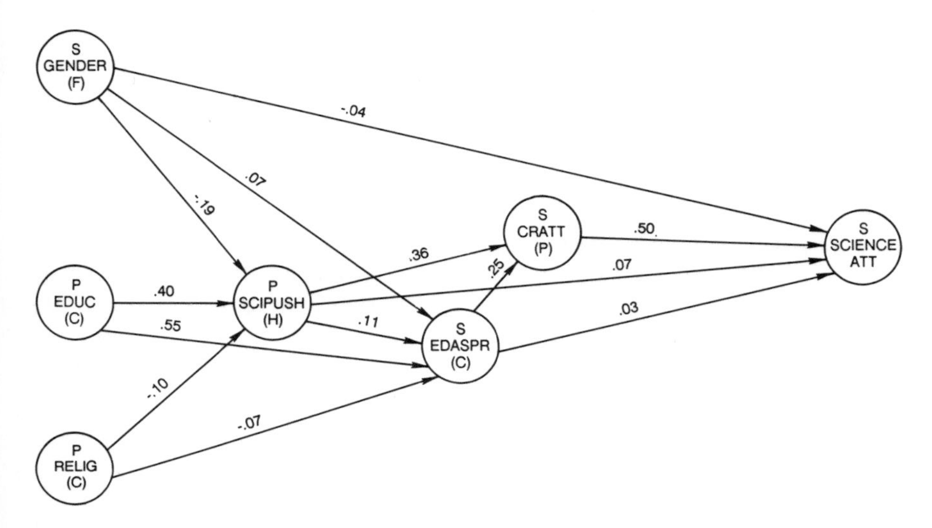

FIGURE 1 **A path model to predict student attitude toward science, 1987.**

sex, parent education, and parent religious attitudes are considered as background variables that have been extant for most, if not all, of each student's life. These three variables are placed on the far left of the model.

Looking at the direct paths, the model indicates that four variables have a direct influence on student attitude toward science. The strongest path comes from student attitude toward his or her science course. The path coefficient is Goodman's (1972) coefficient of multiple-partial determination (CMPD), and the value of .50 indicates that 50% of the mutual dependence in the direct model can be attributed to student course attitude. (CMPD is a proportional reduction-of-error measure. The CMPD uses the difference between the number of likelihood-ratio chi-squares in the independence model and any other model to measure the improvement in estimation attributable to any given model. Goodman suggests that the CMPD is analogous to a multiple R^2 in a regression model. Goodman suggests that in the analysis of ordinal and nominal variables, it is preferable to use the term "mutual dependence" to refer to the deviation from true independence. When two variables are unrelated, we refer to them as being independent. When two variables are not independent, Goodman refers to them as being mutually dependent and measures this mutual dependence in units of likelihood-ratio chi-squares.) In contrast, parent science push accounts for only 7% of the mutual dependence in the direct model. Student sex and student educational aspiration account for even smaller portions of the mutual dependence in the direct model. Parent education and parent religious attitudes do not have a direct path to student general attitude toward science.

It is important to understand that the direct paths are residual paths, in that these relationships express the influence of each of the independent variables on the dependent variable, holding constant the direct and indirect influence of all the other independent variables included in this analysis. The best way to understand this point is to look at an example.

A review of the influence of parent education will be helpful. Beginning at the left side of the model, the path coefficient of .40 between parent education and parent science push indicates that college-educated parents are significantly more likely to push science with their children than are non-college-educated parents. Similarly, the coefficient of .55 between parent education and student educational aspiration indicates that the children of college-educated parents are significantly more likely to plan to earn a baccalaureate than are the children of other parents. In short, better-educated parents foster higher educational aspirations and push science with their children, and the sizes of the two coefficients indicate that both these relationships are strong.

Following this network of influence, parent science push is strongly associated with student course attitude, accounting for 36% of the mutual dependence in the prediction of student course attitude. Parent science push is significantly associated with the level of student educational aspiration, but accounts for only 11% of the mutual dependence in predicting student educational aspiration. The level of student educational aspiration is associated with student course attitude, accounting for a quarter of the mutual dependence in the prediction of student course attitude, which we noted earlier is the strong direct predictor of student general attitude toward science.

Looking at the whole network of direct and indirect influences, it is clear that the level of parent education does play a significant—but indirect—role in influencing student general attitude toward science. The influence of parental education in fostering higher educational aspirations and in pushing science creates attitudes and goals that are conducive to liking science courses, which, in turn, appears to be strongly associated with a positive general attitude toward science. The linkage is indirect, but important.

Although the residual direct influence of student sex was small, an examination of the indirect paths indicates substantially greater influence. The path coefficient of −.19 between student sex and parent science push indicates that male students are more likely to have reported that their parents engaged in science push activities than are female students. In subsequent analyses of the parent interviews, we will explore the parent reports of science pushing activities, but for this analysis, these results indicate that student sex accounts for about 19% of the mutual dependence in the level of parent science push, with sophomore boys reporting the higher level of parent science pushing activities. In contrast, the coefficient of .07 between student sex and student educational aspirations indicates that sophomore girls are more likely to plan to earn a baccalaureate than sophomore boys, accounting for about 7% of the mutual dependence in the prediction of student educational expectations. The absence of a path between student sex and student course attitude means that there was not a significant difference in science course attitudes between sophomore boys and girls, holding constant parent education, parent religious attitude, parent science push, and student educational aspiration.

The path mode indicates that there are significant, but weak, relationships among parent religious attitude, parent science push, and student educational aspirations. The −.10 coefficient between parent religious attitude and parent science push means that parents who hold conservative religious views are slightly less likely to push science than are parents holding moderate or liberal religious views. The −.07 coefficient between parent religious attitude and student educational aspiration means that the students of parents with conservative religious views are slightly less likely to plan to complete a baccalaureate than other students, holding constant parent education, student sex, and parent science push. The absence of direct paths between parent religious attitude and either student science course attitude or student general attitude toward science indicates that there is not a residual direct effect of parent religious views on either of those variables. Given the sizes of the coefficients, the indirect influence of parent religious attitude on student general attitude toward science is very small.

Finally, the model suggests that parent science push plays an important role in fostering positive student attitude toward science courses and science generally. Parent science push is the strongest predictor of positive student science course attitude in this model, accounting for 36% of the mutual dependence in the prediction of student science course attitude.

CONCLUSIONS AND RECOMMENDATIONS

Returning to our original concern about the attitudes of students toward science and the failure of American high-school students to enroll in advanced science and mathematics classes in adequate numbers, this analysis of the data from the Longitudinal Study of American Youth indicates that students' attitude toward their science course is the most proximate and most important short-term influence on more general student attitudes toward science. Since most high-school students experience a biology course early in their high-school program, one important impact of biology courses is that on student attitudes toward science generally. In subsequent analyses, we will explore in greater detail which facets of the science-course experience appear to have the greatest impact on general attitudes toward science, but from the analysis reported above it is clear that the experience of the sophomore student in his or her science class has a significant influence in the more general attitudes of the student toward science.

In addition to the influence of science courses, the analysis indicated that parent science push and student educational aspirations also have significant, but far weaker, influences on students' general attitude toward science. The level of parent education has a substantial indirect influence on the later formation of student attitude toward science. Parent religious attitude has little net influence on students' general attitude toward science. Student sex appears to have mixed effects. Sophomore girls are significantly more likely to plan to earn a baccalaureate than sophomore boys, but boys report greater parent science push than girls.

It is clear that both classrooms and parents play important, but different, roles in stimulating positive general attitudes toward science. Parent education and parent science push clearly contribute to holding higher educational aspirations and to liking high-school science courses. These factors, in turn, appear to foster more positive general student attitudes toward science.

For the purpose of this analysis, classrooms and parents were treated as two separate variables. Unfortunately, in practice, they also appear to function relatively independently. This analysis suggests that one approach to increasing student interest in science might be an increased parental involvement in the science program. Some parents already push science with their children. Other parents may wish to encourage their students in science, but lack the educational background or self-confidence to do so. Increased parental involvement in high-school science programs should be explored as one avenue to focusing and using parental influence in the most productive manner possible.

REFERENCES

Fienberg, S. E. 1980. The Analysis of Cross-classified Categorical Data. 2nded. Cambridge, Mass.: MIT Press.

Goodman, L. A. 1972. A general model for the analysis of surveys. Amer. J. Sociol. 77:1035-1086.

Goodman, L. A. 1978. Analyzing Qualitative/Categorical Data: Log-Linear Models and Latent Structure Analysis. Cambridge, Mass.: Abt Books.

Miller, J. D., R. W. Suchner, and A. Voelker. 1980. Citizenship in an Age of Science. New York: Pergamon Press.

11
What High-School Juniors Know About Biology: Perspectives from NAEP, the Nation's Report Card

INA V. S. MULLIS

LEVELS OF PROFICIENCY

Since 1969, the National Assessment of Educational Progress (NAEP) has been conducting regular assessments of student achievement in a variety of school subjects. As part of its most recent science assessment, in 1986, NAEP assessed the science proficiency of a nationally representative sample of eleventh-grade students composed of 11,744 respondents. The assessment included multiple-choice and open-ended questions about their knowledge, skills, and understanding in four science content areas—the life sciences (biology), physics, chemistry, and earth and space sciences—and their grasp of the nature of science (National Assessment of Educational Progress, 1987). The assessment was also conducted at grades 3 and 7 and was designed to monitor trends in achievement. A comprehensive report of the results is contained in *The Science Report Card* (Mullis and Jenkins, 1988).

The data were analyzed with Item Response Theory scaling techniques and summarized on a composite science scale ranging from 0 to 500 (Beaton

Ina V. S. Mullis, deputy director of the National Assessment of Educational Progress (NAEP), The Nation's Report Card, was a coauthor of *The Science Report Card: Elements of Risk and Recovery*. She was principal investigator for NAEP's study of higher-order thinking skills assessment techniques in science and mathematics and is serving on the advisory board for assessment for the National Center for Improving Science Education.

TABLE 1 Have You Taken Biology?

Response	Percentage	Science Proficiency[a]
Yes	89	296 (1.0)
No	11	268 (1.8)

[a]Jackknifed standard errors are in parentheses.

et al., 1988; Mullis and Jenkins, 1988). Eighty-nine percent of the eleventh-graders reported having taken biology, and their average science proficiency was substantially higher than that of students who had not taken the course (see Table 1).

In addition to average science proficiency, to provide a basis for interpreting the results on the NAEP scale, NAEP defined science proficiency at five levels on the science scale. To characterize these levels, science specialists analyzed the types of items that discriminated between adjacent performance levels on the NAEP science scale and described the skills held by students performing at five anchor points (150, 200, 250, 300, and 350).

Table 2 provides a brief characterization of performance at each anchor point and gives the percentage of high-school juniors performing at or above each level. Virtually all eleventh-grade students performed at or above Level 200, indicating an understanding of simple scientific principles. In the area of biology, these students displayed a rudimentary knowledge of the structure and function of plants and animals. They are likely to recognize the characteristics of common aquatic birds and know that the wolf and dog are closely related, that a mouse does not lay eggs, and that the main function of the heart is to pump blood to all parts of the body. Also, as typified by Level 250, most (85%) are likely to be familiar with food chains, to understand that light and water affect plant growth, and to be able to identify how some diseases are transmitted.

While most high-school juniors attained the three lowest proficiency levels, fewer than half reached Level 300—a level characterized by more specific scientific knowledge and the ability to analyze scientific procedures and data. Further, only 6% of the students at this grade level demonstrated the ability to infer relationships and draw conclusions using detailed scientific knowledge.

In addition to highlighting the distribution of students across the five levels, Table 2 reveals large performance gaps between males and females and particularly between white students and their black and Hispanic peers. For example, half the males reached Level 300 in 1986, compared with only

TABLE 2 Percentage of Eleventh-Grade Students At or Above Five Science Proficiency Levels: 1986[a]

Level	Description	Total	Male	Female	White	Black	Hispanic
350	Integrates specialized scientific information	6 (0.5)	9 (0.7)	3 (0.3)	7 (0.5)	--	1 (0.2)
300	Analyzes scientific procedures and data	42 (1.2)	50 (1.2)	33 (1.2)	50 (1.2)	9 (1.3)	14 (1.9)
250	Applies basic scientific information	85 (0.5)	89 (0.5)	82 (0.7)	93 (0.4)	52 (1.7)	66 (2.1)
200	Understands simple scientific principles	99 (0.1)	100 (0.1)	99 (0.2)	100 (0.1)	96 (0.5)	98 (0.4)
150	Knows everyday science facts	100 (0.0)	100 (0.0)	100 (0.0)	100 (0.0)	100 (0.0)	100 (0.0)

[a]Jackknifed standard errors in parentheses.

one-third of the females. And, while half the white students reached this level, only about 9% of the black students and 14% of the Hispanic students did so. Although 93% of the white students reached Level 250, only about half the black and two-thirds of the Hispanic eleventh-graders did.

RESULTS BY SEX AND RACE-ETHNICITY ON THE LIFE-SCIENCES SUBSCALE

To construct the composite science scale, NAEP computed results for the different content-area subscales—one of which was life sciences (biology). The meaning of the science subscales cannot be known in absolute terms; that is, one cannot determine how much learning in chemistry equals how much learning in the life sciences. The subscales do, however, permit an analysis of the relative strengths and weaknesses of students in different population groups within each science content area.

Like the proficiency results on the composite science scale, the results on the life-sciences subscale indicated that white eleventh-graders outperformed Hispanic eleventh-graders, who outperformed their black counterparts. Given these results, a natural question revolves around the impact of any potential differences in biology course-taking for these groups of students. However, as shown in Table 3, although slightly fewer Hispanic students than white or black students reported having taken biology, the percentages were equivalent for white and black students (Mullis and Jenkins, 1988). In all three racial-ethnic groups, students who had taken the course performed much better on the life-sciences subscale than those who had not, but course-taking did little to lessen the performance gaps between these groups of students. The gaps remained essentially constant.

Also in keeping with results on the composite science scale, males performed better than females on the life-sciences subscale, although it is

TABLE 3 Average Proficiency of Eleventh-Grade Students on the Life-Sciences Subscale, by Biology Course-Taking[a]

Student Group	Percentage of Students Who Have Taken Biology	Proficiency of Students Who Have Taken Biology	Proficiency of Students Who Have Not Taken biology
Male	88 (1.0)	298 (1.3)	271 (3.0)
Female	89 (0.9)	291 (0.9)	265 (2.5)
White	89 (1.0)	302 (0.8)	276 (2.5)
Hispanic	82 (2.3)	269 (1.6)	248 (2.4)
Black	89 (1.3)	259 (1.7)	239 (3.0)

[a]Jackknifed standard errors are in parentheses.

interesting that this difference did not appear to be substantial until grade 11. In fact, at grade 3, girls had a slight edge in performance—an edge that shifted in favor of boys by grade 7.

Table 3 also indicates that just as many females as males take biology in high school. However, while both male and female students who had completed biology performed significantly better on the life-sciences subscale than those who had not taken the course, sex differences in performance remained essentially unchanged, irrespective of biology course-taking.

Patterns of high-school performance of groups of students defined by sex or race-ethnicity appear to be established early in the schooling process. For example, although the sex gap was not evident in the area of biology for third-graders, there was a substantial discrepancy in performance among racial-ethnic groups, with white students having significantly higher proficiency levels on the life-sciences subscale than their black and Hispanic classmates.

Forty-six percent of the seventh-grade students described life science as the primary area of study in their science classes. However, these students did not perform as well on the life-sciences subscale as the one-quarter of their classmates who reported studying general science, although students studying life science did outperform the 6% of the students who reported study emphasizing physical science and the 11% studying earth science. By seventh grade, for both students studying life science and those studying general science, performance gaps were apparent between males and females, with the males having the higher levels of achievement. The gaps among the three racial-ethnic groups were substantial, with white students outperforming their black and Hispanic peers.

SELECTED ITEM-BY-ITEM RESULTS

At grade 11, 59 items were included in the life-sciences subscale, and the performance results for some of these items are provided below to illustrate the composition of this subscale and to present some specifics about what high-school juniors know about biology. The items are categorized in four groups for the purposes of discussion: ecological relationships, cell structures and functions, energy transformation, and genetics.

It should be emphasized that the items that follow are only illustrative of the skills, knowledge, and understanding tapped in the 1986 science assessment and are not intended to be an inclusive account of all that high-school students should know about or be capable of doing in biology (Mullis and Jenkins, 1988).

Ecological Relationships

Eleventh-grade students appeared to have some knowledge of their environment and a grasp of basic ecological concepts. For example, in response to the item below, 81% of the students correctly identified the fox as the predator in the food web presented.

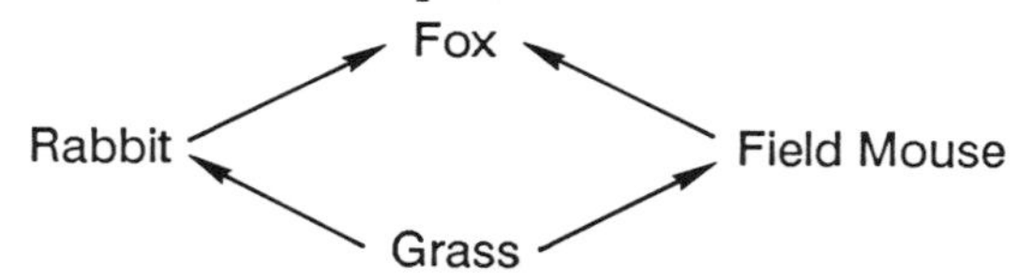

With respect to the field mouse in the food web above, what is the fox considered?

- • A predator
- o A prey
- o A producer
- o A decomposer

In response to several other ecology-related items, 80% of the high-school juniors recognized acid rain as a kind of pollution, and 77% knew something about the effects of insecticides. In contrast, only about one-third (31%) of the students recognized a recommended method for controlling soil erosion, and fewer than one-fifth (19%) correctly identified a graph of the world's population growth.

For the items described above, the response patterns for groups were similar to those shown on the life-sciences subscale overall: eleventh-grade males performed significantly better than females, and white students performed significantly better than either black or Hispanic students.

Cell Structures and Functions

Although two-thirds of the eleventh-graders recognized a diagram of a group of cells as a tissue, only one-third were able to identify the basic function of the cell membrane. Fewer still (approximately one-fourth) were able to apply their knowledge to specify how a cell membrane works or explain the distinguishing features of a plant cell based on a diagram. Although white students tended to perform better than black or Hispanic students on these questions on cell structures and functions, the differences in performance between males and females were minimal.

Energy Transformation

While 90% of the high-school juniors recognized that "junk food" is high in calories and low in nutrients, fewer than two-thirds appeared to have a general understanding of photosynthesis. For example, 60% of the students responded correctly to the item below.

Which of the following best explains why marine algae are most often restricted to the top 100 meters in the ocean?

- o They have no roots to anchor them to the ocean floor.
- • They are photosynthetic and can live only where there is light.
- o The pressure is too great for them to survive below 100 meters.
- o The temperature of the top 100 meters of the ocean is ideal for them.

A slightly higher percentage of the students appeared to know that plants produce oxygen and tend to grow toward light (68% and 67%, respectively). Somewhat fewer students (57%) were able to apply their knowledge to respond correctly to an item on the role of red blood cells in transporting oxygen.

In this area, again, the sex gap was not significant, perhaps because females were more likely than males to report experience in working with plants and animals. As with performance on the groups of items previously described, however, there were significant differences in the performance of white, black, and Hispanic students on items pertaining to cell structures and functions.

Genetics

As made evident by their performance on items in this category, genetics was a relatively difficult area for the eleventh-grade students assessed. Approximately half the students demonstrated a basic understanding of recessive genes, and 57% could identify the probability that parents with a certain genetic structure would produce blue-eyed children. Forty-seven percent of the students responded correctly to the item below.

A female white rabbit and a male black rabbit mate and have a large number of baby rabbits. About half of the baby rabbits are black, and the other half are white. If black fur is the dominant color in rabbits, how can the appearance of white baby rabbits best be explained?

- o The female rabbit has one gene for black fur and one gene for white fur.
- • The male rabbit has one gene for black fur and one gene for white fur.
- o The white baby rabbits received no genes for fur color from the father.
- o The white baby rabbits are result of accidental mutations.

Fewer students (28%) were able to use their knowledge of natural resistance to assess the implications of genetics research, as shown in the example below. Although males and females performed similarly on questions related to genetics, white students tended to outperform their black and Hispanic classmates.

Recombinant DNA research has produced a variety of organisms with big economic potential. For which of the following reasons are concerned citizens hesitant to permit the use of these organisms outside of the laboratory?

- o Production of such organisms will involve the production of hazardous by-products.
- o Most scientific research is perceived to be dangerous.
- o The organisms could die outside of a laboratory environment.
- • The introduction of organisms new to the Earth could upset the ecological balance.

SUMMARY

Nearly all (89%) of the approximately 12,000 high-school juniors assessed by NAEP in 1986 reported having taken a course in biology. Students who had taken the course performed significantly better than those who had not on both the life-sciences subscale and the composite scale representing performance in all the science content areas assessed.

Given that most high-school juniors have taken biology, their understanding of the life sciences appears quite limited. Virtually all these students exhibited the kinds of knowledge that may be gained from everyday experiences; however, substantially fewer displayed more detailed knowledge and understanding. While an instructional emphasis on some aspects of ecology is suggested in the performance results, students appeared to have had relatively limited or ineffective exposure to other areas in the life sciences. For example, students displayed little understanding of cell structures and functions, genetics, and energy transformation.

On the basis of their lack of knowledge, skills, and understanding and their inability to apply those they do possess, it is likely that our high-school juniors do not grasp the larger concepts that most science educators believe to be the foundation of a strong education in biology, including systems and cycles of change, heredity, diversity, evolution, structure and function, and organization.

These findings, while troubling in themselves, are given further weight by evidence that substantial disparities exist in the performance of groups defined by sex and by race-ethnicity. While just as many females as males had taken biology, course-taking in this subject did not appear to lessen the sex-related performance gap. Rather, the gap remained as large among students who had taken the course as among those who had not. Similar patterns were found in the disparities across racial-ethnic groups. For each of the three groups analyzed by NAEP—white, black, and Hispanic students—students who had taken biology outperformed those who had not taken the course. However, biology course-taking did not appear to lessen the performance gaps between white students and their black and Hispanic peers.

REFERENCES

Beaton, A. E., et al. 1988. Expanding the New Design: The NAEP 1985-86 Technical Report. Princeton, N.J.: Educational Testing Service.

Mullis, I. V. S., and L. Jenkins. 1988. The Science Report Card: Elements of Risk and Recovery. Princeton, N.J.: National Assessment of Educational Progress, Educational Testing Service.

National Assessment of Educational Progress. 1987. Science Objectives: 1985-86 Assessment. Princeton, N.J.: National Assessment of Educational Progress, Educational Testing Service.

12

The NABT-NSTA High-School Biology Examination: Its Design and Rationale

BARBARA SCHULZ

EVOLUTION OF THE BIOLOGY TEST

The evolution of the national high-school biology test is especially interesting. Early in this decade, there was much discussion about the need for such a test by both the National Science Teachers Association (NSTA) and the National Association of Biology Teachers (NABT). Each group initiated preliminary discussions about whether such a test should be written, how it would be written, who would write it, and how it might be abused. In 1982, the high-school division of NSTA, under the direction of Linda Perez of Texas and Angelina Romano of New Jersey, conducted a needs assessment among the membership. The response was overwhelmingly in favor of a test. At the same time, NABT appointed a small committee to discuss the feasibility of such a test. This group, including Joe McInerney of Colorado and Ken Bingman of Kansas, discussed the notion of developing a test bank of questions available to the membership. In 1984, the NSTA board of directors passed a motion that NSTA and NABT proceed with the development of a national test. The motion also asked that the president

Barbara Schulz has been a high-school science teacher for 22 years and department chair since 1974. She was a recipient of the Outstanding Biology Teacher Award in 1981 and the Presidential Award for Excellence in Science Teaching in 1983, and she was a semifinalist for the Teacher in Space Award. Ms. Schulz was president of the Washington Science Teachers Association in 1987-1988.

of each organization appoint four persons to serve on a joint test development committee. So it was that the first NABT-NSTA test development committee met in Chicago in June 1985 to begin the test construction.

It was with a great deal of hesitation that this joint committee of nine biology educators proceeded with the test development. Some states have competence tests; others are considering such a move. We, the professional biology teachers, feel best qualified to design the test and help to set the direction of the biology curriculum. With the great demand for accountability being felt by all, it was clear that a test would be developed. The following statement of rationale and purpose was prepared and printed in both *News & Views* and *The Science Teacher Journal* in 1985:

A Standardized Test for First-Year High School Biology
Rationale and Statement of Purpose

There is an increasing demand for accountability in science education, and science educators, through their professional organizations, should assume responsibility for establishing the mechanisms for that accountability, lest the responsibility fall to lay persons with vested interests. One such mechanism is a student education instrument. Accordingly, the National Association of Biology Teachers and the National Science Teachers Association are collaborating on a project to develop a standardized test for high school biology. This objective, year-end test will be intended for first-year high school biology students and will address a core of basic biological concepts, processes, and thinking skills. The joint committee has agreed on the following principles:

a. The test should be used to improve science education; questions will be oriented toward inquiry and other higher-level cognitive functions;
b. The test should not be used to evaluate teachers;
c. The test should not become an end in itself, that is, the biologic content reflected in the test items should not be interpreted as the final word on a complete conceptual framework for an introductory biology course; and
d. The test should be updated every two years.

VALIDATION BY THE MEMBERSHIP

Having declared that the purpose of this test is to drive curriculum forward, the committee looked at the question of content and level of difficulty. The following concepts were decided on:

A. Cell structure and function
Sample concept: Biological systems vary in their degree of specialization.*
B. Bioenergetics
Sample concept: Biological systems cannot exist without energy input.*

*Biology Assessment Review Workshop. Biology Test Domains, Objectives and Content Specifications. Wisconsin Department of Public Instruction. 1983.

C. Genetics
Sample concept: Organisms pass on characteristics to the next generation through genetic material.*
D. Evolution
Sample concept: Organisms change through time.
E. Systems, physiology to morphology
Sample concept: Structure and function complement each other in biological systems.
F. Ecology
Sample concept: Organisms are interdependent, and their interactions result in the flow of energy and the cycling of matter.
G. Taxonomy
Sample concept: Biological systems are grouped on the basis of similarities that reflect evolutionary history.*
H. Behavior
Sample concept: The response of an organism to its environment has both a genetic and an environmental basis.
I. Science, technology, and society
Sample concept: Advances in science and technology have implications for personal and societal decision-making.

It was also decided that the following list of processes and skills should be represented in the way questions were designed:

1. Inquiry
2. Process science
3. Experimental design
4. "Science as a way of knowing" (John A. Moore)
5. History
6. Probabilistic thinking
7. Creative problem-solving

These concepts, processes, and skills were published in the same article with the rationale. A response card was published concurrently in both journals, and readers were asked to validate the conceptual framework and intent of the test. The test committee received a good response; more than 400 cards were returned. Also, sessions were held at the three NSTA regional meetings, the NSTA national convention, and the NABT national convention for the purpose of concept validation. As a result of the meetings, the conceptual area of science, technology, and society was added to the test by popular demand. A call for test questions from the membership was made and a good response received. Each question was reviewed by a college-level content specialist and a high-school biology teacher for validity and appropriateness. From the solicited questions, 120 were selected for review and field-testing. Of those, 80 were selected for the final test.

TEST CONSTRUCTION

Armed with more than 300 questions, the committee turned to the issue of test content. The following numbers of questions were agreed on:

Concept	No. questions
Cell structure and function	8
Bioenergetics	10
Genetics	12
Evolution	12
Systems, physiology to morphology	8
Ecology	8
Taxonomy	6
Behavior	8
Science, technology, and society	8

In addition, the committee decided that each concept area should be written at three levels of difficulty, ranging from knowledge to synthesis.

A field test of 120 items was given in the spring of 1986 to students in 12 states covering all regions of the country. The Lertap test analysis was done on the field-test data. Eighty questions were then selected for inclusion in the first edition of the test. A second field test was done on the 80-item test in the fall of 1986 after some editing of the positively correlated distractors.

THE RESULTS ARE IN

By the spring of 1987, the test was published, and more than 30,000 copies were sold. The Educational Testing Service (ETS), Princeton, New Jersey, was hired to do the data analysis on the first commercial edition of the test. ETS analyzed 895 answer sheets from Form A and 1,075 answer sheets from Form B.

	Form A	*Form B*
Mean	43.8	40.7
Standard deviation	13.3	12.4
Median	44.3	40.2
Reliability (alpha)	0.91	0.91

Committee member Juliana Texley provided the analysis shown in Figures 1, 2, and 3.

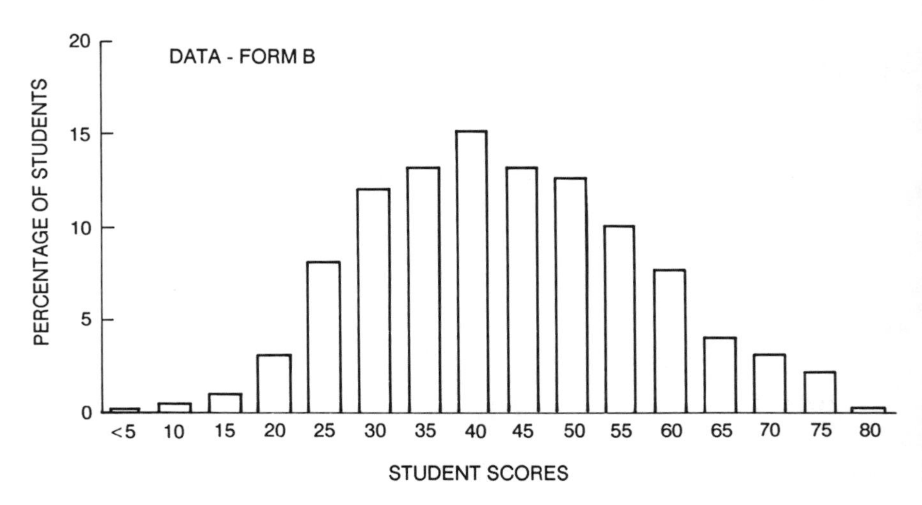

FIGURE 1 Test Form B.

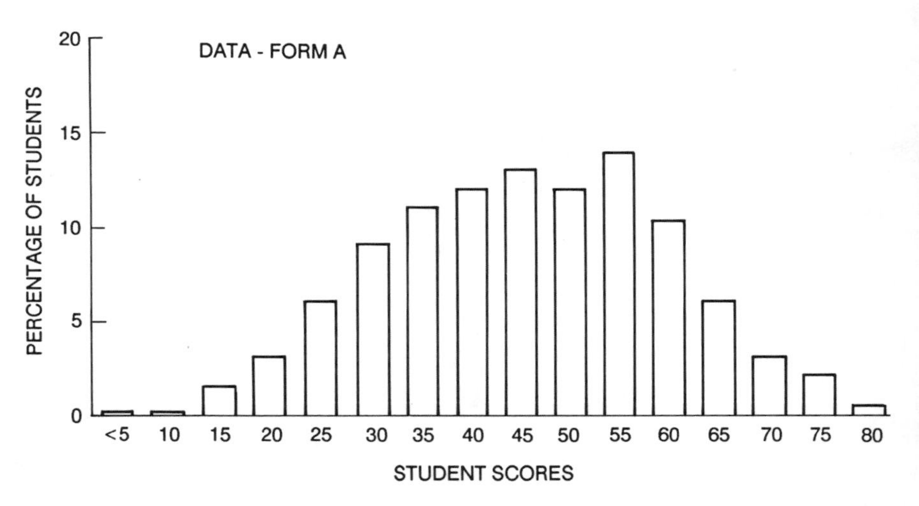

FIGURE 2 Test Form A.

The test committee was pleased with the results. There was a strong interest by science teachers in using this test, and we have learned much from it. The committee made some minor revisions to correct misleading language in the 1987 test. The revised edition will be available for the next 2 years. In 1990, at the request of the memberships of NABT and NSTA, a completely new test will be designed and tested. The same format will be followed.

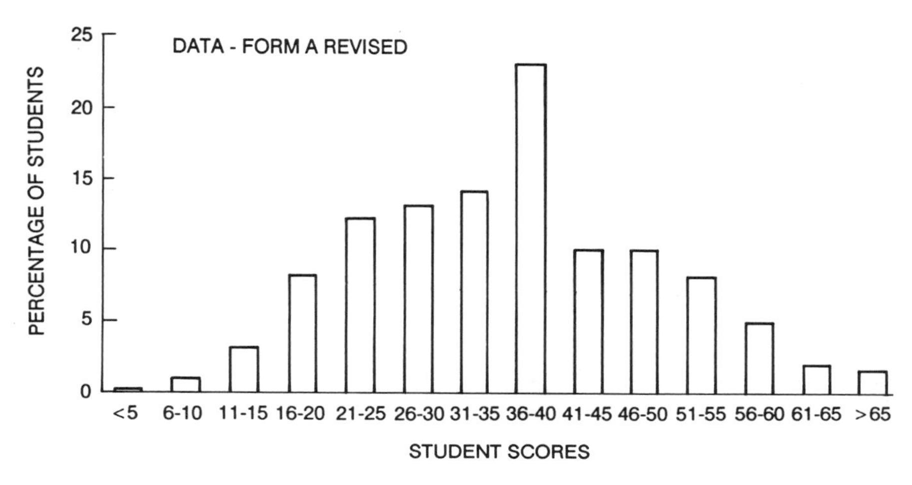

FIGURE 3 Test Form A Revised.

CONTINUED CHANGE THROUGH TIME

The test will be recreated every 2 years. New committee members will rotate on to the test committee to serve a 3-year term, the time it takes to build, field-test, and publish the test. Questions will continuously be sought from practicing high-school biology teachers. Thus, the curricular emphasis within the test can change to reflect the concepts deemed relevant by biology teachers. The test may help to focus a solid core curriculum in biology, stated in terms of broad and significant concepts, rather than encyclopedic facts.

The biology test, as an effort of professional associations, is not a product, but an on-going process. The first administration of the test has generated at least as many questions for the committee as we provided for our students. As the data from the first test were analyzed, the group was already planning a schedule for soliciting and field-testing new items for future tests. The committees will change, and the evaluation will evolve with the help and input of members. And as they do, we will learn more about our students, our classrooms, and ourselves.

After 2 years of development, the first test results for biology are in, and another benefit of the test development process has also surfaced. We quickly realized that we have been given a valuable insight into what students know—about the science of biology. Scanning the answers of over 2,000 randomly selected subjects, the committee was able to peek into some widely held misconceptions and to hypothesize about the classroom procedures that might be perpetuating them. To a large extent, the test's validity derives from the many years of joint experience the committee members share; their interpretations of the data they have received from the first test administration come largely from the same experience base. The group hopes to engender further discussion to improve not only the test, but the process of biology education.

A PEEK INTO STUDENTS' MINDS

Every teacher knows that the hardest part of test construction is choosing the wrong answers (the distractors or foils)—not too easy to spot, not so outlandish that no one would choose them. When the constructors of the national biology test received their item analyses, one of their most significant data sets was the percentage of students that chose each of the wrong answers. When one foil attracted a very large number of respondents, the obvious question always came up: Why? In a few cases, the foil was found to be marginally correct, given a slightly nontraditional reading, in a way the group had not foreseen. This type was changed by editing. But there was nothing "right" about many of the most frequently chosen foils. What was happening, it seemed, was that the foil touched on a widespread student misconception or a teaching technique that often misfired.

Some of the common errors seemed mnemonic; they seemed to result from verbal associations that we repeat too often in the teaching of biology:

- When we asked students to complete the phrase. "The cell is a unit of structure and a unit of _____," in the first field test, we were amazed to find that the majority chose "organ system." Had we taught the sequence "cell, tissue, organ, system" by rote once too often?
- Students demonstrated common vocabulary confusions, such as mistaking "cell membrane" and "cell wall."
- When we asked a question about meiosis, the most popular choice was one that contained the word "gamete"—even though the sense of the answer was completely incorrect.

The results suggested to the committee that far too many of our students are relying on word associations to weave their way through biology. Do such tricks work on classroom tests? Do we encourage them?

Other common errors that the students demonstrated told us that some of our most important conceptual goals were often not met. In questions about evolution, the Lamarckian explanation for an adaptation was consistently chosen as often as the explanation based on natural selection. This result is backed up by a number of research studies that show that the idea of inheritance of acquired characteristics is both intuitively appealing and surprisingly persistent in biology students of all ages.

Similarly, the concepts of energy and entropy were difficult for students, despite the relatively simple and straightforward wording of questions. The idea that energy dissipates and does not cycle in the environment was a difficult one for students in several contexts. Perhaps too much emphasis was placed on cycles and too little on energy.

We saw evidence of misconceptions that were and are text-perpetuated. Students believe (on the basis of misinformation in many texts) that mutations are always recessive and weak (Mahadeva and Randerson, 1982).

We also found that it was dangerous to assume that students had experienced some of the more common laboratory investigations in first-year biology texts. Students found two questions about surface-to-volume ratio very difficult; it seemed that they had not explored the relationship between cell-membrane size and cell division. Any questions about hypertonic and hypotonic solutions were quite challenging when the terms themselves were not used. It seemed that students relied on the words, rather than the experiences, to influence their predictions.

OTHER VARIABLES IN THE TESTING PROCESS

In analyzing an evaluation tool, test-makers must be conscious of the other factors that can contribute to the variance in student performance. In the construction of the national biology test, the authors paid careful attention to the reading level and vocabulary of each question. In many cases, judicious editing was effective. But there was still evidence that the longer questions were harder than the shorter ones—a result that was not expected.

What was surprising, and what may provide the basis for more detailed research by the group, was evidence suggesting that questions involving visual or graphic analysis were harder as a group than the others in the instrument.

The students who took the test seemed consistently confused by graphs and diagrams. In one item set based on an enzyme graph, the independent variable was increasing left to right, but students commonly erred by assuming that the enzymes represented left to right were in the order of their presence in the alimentary canal; that is, many believed the first were mouth enzymes, the second stomach enzymes. In a diagram of predator-prey relationships in a prairie, many students guessed that the prey of coyotes in that community would be jackrabbits—despite the evidence provided in a graph and clear directions to answer the question from that graph. In 1988, the committee examined seven demographic questions relevant to performance on the high- school biology examination. Each of the questions was precoded onto the test forms by students in self-selected classrooms and analyzed by means of one-way analysis of variance at a 0.05 level of significance. We randomly selected 882 tests. Although Forms A and B were distributed that year, only Form A responses were available in numbers suitable for random selection for analysis. (Previous analysis of test scores indicated that the forms were parallel, since only the order of the answers had been changed.)

Of the students in the sample, ninth-graders and eleventh-graders did significantly better than the tenth-graders who would normally be enrolled in standard-level biology classrooms.

In analyzing the data further, we found that students who indicated that they "never" experienced laboratory work did significantly more poorly than those who did laboratory work "some of the time." The frequency of laboratory work was not an important factor. However, those who had a laboratory experience did better than those with no laboratory experience or those who said they had laboratory all the time. While this identifies laboratory experience as necessary, it also brings into question student perception of "seldom," "frequent," and "most of the time."

There is little evidence of standardization among advanced-biology sections, and some of these students may have been in courses tailored to individual research. The committee found no significant difference based on structure of schools. However, there was some evidence that students in smaller schools—500 or fewer—performed significantly better.

Our results on item difficulty gave us a clue to what was and what was not generally taught in the classrooms where our normative data were developed. Botany questions were uniformly more difficult for students than zoology questions. Mendelian genetics was surprisingly easy; modern genetic engineering was often very difficult. Taxonomy questions were the easiest (even though the test did not ask any specific taxa). And questions about the societal implications of modern biology and environmental problems like acid rain were answered correctly by very few subjects, suggesting that teachers may be reluctant to add this emphasis to their curriculum.

FUTURE TESTING

For the immediate future, the committee has opted to add clearer pictures and diagrams for students who need such help. In years to come, both teaching and testing may be enhanced by far more visual stimuli; videotape and real-life examples may help students to reason more effectively with broader comprehension.

Perhaps the most important implication of such a national test is not the result, but the point from which the committee started. With the recognition that we can't teach—and students can't really learn—everything in the commercial texts, the joint position of the associations is that the test establishes a core of nine concepts that should be a part of every student's first-year biology experience. It was this list—and not the questions themselves—that seemed to elicit the most interest in the members, many of whom would rely on such a statement to guide their own choices.

REFERENCE

Mahadeva, M., and S. Randerson. 1982. Mutation mumbo jumbo. Sci. Teach. 49(3):34-38.

PART III

Curriculum: Perspectives and Content

13

The Evolution of Biology and Adaptation of the Curriculum

TIMOTHY H. GOLDSMITH

At the outset I would like to salute those many dedicated high-school teachers who are doing a marvelous job under far from ideal circumstances. They are true professionals, and continuing to nurture and support them is one of the challenges that face us. But it is *despite* their best efforts that this conference is being held.

In little more than a century, the science of biology has undergone two "evolutionary" changes of major magnitude. First, of course, was appreciation of the reality of organic evolution and its power as an explanatory principle, a change that only began with Charles Darwin. Second was insight into the structure of the genetic material, DNA, which opened the way to the broad range of both techniques and fundamental understanding of basic biological processes that are encompassed by the term "molecular biology." The first of these events provided a new and profoundly important way to view the natural world. The second has led to such enormous progress that virtually for the first time in the history of our science we can ask meaningful experimental questions about such central problems as how a fertilized egg develops into a functional adult organism and how a collection of neurons can learn and remember.

I would like to set the stage for this session on perspectives and curricular content by stating a proposition, perhaps audacious, but one

---

Timothy H. Goldsmith, a neurobiologist, is professor of biology at Yale University. He is a member of the National Research Council's Board on Biology.

I believe to be defensible. Not just despite, but in some sense *because* of, these exciting changes in biology, our educational system has failed in deeply important ways. Not for a total want of trying. There have been commendable and temporarily or locally effective efforts, of which the Biological Sciences Curriculum Study project is the most noteworthy. But viewed over time, instead of a harmonious coadaptation of the biology curriculum and the science of biology, we see episodic outbursts of interest, followed by periods in which—in my metaphor—selection is relaxed. We forget that evolution is unremitting change.

I see no blanket prescription for dealing with this dilemma, for it represents a complex of problems. But let me try to focus on several that have to do with our theme. I am not going to offer solutions, for my present role is to learn. But I am going to point to some of the broader issues that lurk in the background, forming part of the social fabric on which we must embroider.

The proper teaching of evolution has not been solved. Our national tradition of local autonomy in education has produced an anomalous situation where perceived local social and religious values determine the content of nationally marketed textbooks and warp the scope of the science curriculum. As in other subjects, we have virtually no national standards in a world of international competition. The situation is so bad, according to a recent study, that 19% of biology teachers believe that humans and dinosaurs lived at the same time. But the problem only starts in the schools. By the time the most talented and motivated students elect to pursue the study of biology further, many of them fail to understand that biological questions always have two kind of answers—one reductionist in nature, the other historical—and that these two quite independent explanatory approaches are of equal intellectual validity and importance.

Evolutionary biology is not stamp-collecting, and understanding biological diversity is an immensely important task. If we view ourselves as part of nature, we are more likely to develop a respect for the only Earth we have, a theme eloquently developed earlier here by John Harte. We may also view our own behavior in new and different lights. At an intellectual level, most of the political arguments that energize democracy reflect philosophical disagreements about the relative importance of different facets of human nature. At a practical level, most political struggles, and the wars they generate, involve competition for resources. One can make the case that the religious and political rationales for conflict are but evocations of group identity to solidify effort in the protection of presumed common interests. What passes for political dialogue is frequently a vocal demonstration of how easy it is for the limbic system to escape control by the cerebral cortex. All of this involves interesting biology, evolutionary biology.

The religious fundamentalists are correct in their expectation that proper education in biology will produce citizens prepared to question many of the traditional assumptions of society. I would firmly disagree, however, that this must undermine the inculcation of moral values. But, frankly, this is not the central issue. As was stated in a recent letter to the editor in *The New York Times*, "Allan Bloom is wrong—there can be no closing of the American mind, for it has never really been opened" (New York Times, 1988a). At the risk of projecting a pessimism I do not in fact feel, this sentiment is an echo of Alexis de Tocqueville (1956), who observed a century and a half ago, "I know of no country in which there is so little independence of mind and real freedom of discussion as in America."

We fare little better in teaching the parts of the science that are related to the new molecular biology, but for rather different reasons. Traditionally, school biology has been offered before chemistry and physics. This has made sense, for it is easier to introduce the unknown by way of the known. Plants and animals are familiar to children; the concepts of atoms and molecules, coulombs and photons are not.

But as the pace of discovery in biology has increased, there has been an understandable wish to bring the latest news to the classroom. My impression, however, is that we are not very clever about teaching biological concepts—many of which have an intrinsic beauty—without either smothering students in the vocabulary of biochemistry at a time when they have little or no idea what it means to be a molecule or confusing them with presentations that have been edited into chaos by people who do not have appropriate knowledge. I have known high-school students who could tell you, haltingly, that DNA stands for deoxyribonucleic acid, but ask them another question about DNA, and you find that you have seen to the horizon of their understanding. At its best, the result of this kind of education is likely to be tedium. At its worst, it provides wrong information. And somewhere in the middle lies confusion.

It is important to recognize the larger context in which we face this problem. It is not just the teaching of biology, or even science, that has this disease. The Bradley Commission on History in the Schools has recently called for more emphasis on broad trends and questions and on the teaching of critical thinking, rather than the memorization of facts without context. Less than 2 weeks ago, Kenneth Jackson, the commission chairman, was quoted as saying that "history should not be just a mad dash through the centuries with teachers trying desperately to get to the 1980s before school lets out in June" (New York Times, 1988b). By changing only three words, that sentence could just as well address the presentation of biology. And that, I submit, may be telling us something important.

Could it be that a citizenry that resonates so easily with the notion that teachers should be required to lead their classes in the Pledge of

Allegiance to the flag is really more interested in an educational system that indoctrinates than an educational system that teaches critical thinking? We should not take refuge in the thought that science, being "objective," is immune to this influence; our experience with the teaching of evolution shows otherwise. No, on this issue we should be making common cause with our thoughtful colleagues in the humanities, for our aspirations for the children of this nation are fundamentally the same.

We need to ask what it is we are trying to do and for whom we are trying to do it. Only when we have answered those questions can we address the specifics. But somewhere in the process we should ask whether we have the right relationship between the sciences in the high-school curriculum. Do we do things in the right order and with the right degree of integration? And if we do not, what must we do to change? What do we need to do to bring observation, excitement, and the joy of discovery to the classroom? And can we hope to inject these same goals into the elementary-school years without measuring our progress on the geologic scale of time?

Finally, I would like to suggest that there is not enough *imagination* in what is taught. All too frequently, pedestrian or muddled presentations of elegant concepts fail to connect with the backgrounds, interests, and needs of the children. It is not the ideas themselves that are inappropriate, but the way they are treated in many of the textbooks. Is it hopeless to expect more of an author-editor formula that appears insensitive to accuracy and nuance and explains material to the student with all the finesse of a delivery of loose gravel? If we as a nation are going to get excellence in education, the textbook industry will have to show more concern for real expertise in both biology and teaching and less of a preoccupation with mass marketing.

I have developed impatience with the assertion that publishers cannot afford to produce material unless it conforms to some lowest common denominator that enables it to be sold nationally. This is not true for college textbooks. I therefore conclude that it is a doubtful proposition in the first place, and one that we have accepted passively for far too long.

I hope I may be persuaded in what is to follow that we are moving in some of the right directions, and that in its own evolution, the curriculum is at last adapting to the needs of both science and society.

REFERENCES

de Tocqueville, A. 1956. Democracy in America, p. 12. Abridged and edited by R. D. Heffner. New York: New American Library.
New York Times. 1988a. Letters to the Editor, Sept. 17.
New York Times. 1988b. Pg. A36, Sept. 30.

14

Human Ecology: Restoring Life to the Biology Curriculum

JOSEPH D. McINERNEY

Anyone who undertakes an examination of the high-school curriculum—irrespective of the subject matter—would do well to consider Garrett Hardin's (1985) first of several "postulates of impotence" that guide ecological thinking: "We can never do merely one thing." The content of the curriculum influences and is influenced by so many variables—from budgets to buses—that to consider the curriculum in isolation is pure folly. Nonetheless, my task is to address the content of the biology curriculum, and that will be my central focus. I shall allude briefly to other issues that are inextricably bound to content, but shall leave the full explication of those issues to others who are more qualified to give them the attention they deserve.

REFORM IN SCIENCE EDUCATION

The 5 years since the publication of *A Nation at Risk* (National Commission on Excellence in Education, 1983) have been interesting, confusing, and sometimes frustrating for those of us who spend our time thinking about and developing science curricula. Since the publication of the report, there

Joseph D. McInerney received his undergraduate degree in education in 1970 from the State University of New York (SUNY), Cortland, and an M.S. in human genetics in 1975 from SUNY, Stony Brook. He joined the Biological Sciences Curriculum Study in 1977 and has been its director since 1985. He is a member of the editorial board of *Quarterly Review of Biology*.

have been more than 100 attempts (Mullis and Jenkins, 1988) to clarify what Americans educated at the high-school level in science should know about science. Project 2061 (Rutherford and Ahlgren, 1988), sponsored by the American Association for the Advancement of Science, is perhaps the most complete, with its emphasis on all disciplines and its suggestions for what should be omitted from the already overcrowded science curriculum.

With respect to biology, the project Science as a Way of Knowing (SAAWOK), organized by the American Society of Zoologists and cosponsored by nine other professional societies, has been particularly informative, notwithstanding that it is intended to induce change in the undergraduate curriculum. SAAWOK has demonstrated anew—as the Biological Sciences Curriculum Study (BSCS) did in the 1960s—that one can take any of several conceptual approaches to biology (such as evolution, human ecology, genetics, development, form, and function) and do a first-rate job of conveying essential, enduring principles of the discipline. Each approach, in fact, can encompass the others.

Given that any of several approaches will convey the principles of the discipline very well, curriculum developers must ask: "Which approach is most likely to meet the educational needs of all high-school students?" That is, what is the proper approach for students who will likely have no further formal exposure to biology, as well as future biologists? This question is very different from one that influenced the reform movement of the 1960s and 1970s: "How can we best prepare young people for careers in biology?" The answer to that question was to develop curricula that focused on the structure of the discipline under consideration (McInerney, 1987). The assumption this time around, however, is that the wave of reform should reach farther up the beach to encompass *all* citizens, not only those who wish a career in science, and not only those whom Jon Miller and co-workers (1980) called "the attentive public for organized science." We must, therefore, take a different view of the curriculum, and there is an emerging consensus that the objective of the science curriculum should be the development of scientific literacy in the general public.

Achieving consensus on the definition of scientific literacy, however, has not been quite so easy. The definition I shall use is taken from a 1983 essay by Kenneth Prewitt; I consider it the best definition of the many I have seen in the current upsurge of interest about science education:

> From the perspective of democratic practice, the notion of scientific literacy does not start with science itself. Rather, it starts at the point of interaction between science and society. My understanding of the scientifically savvy citizen . . . is a person who understands how science and technology impinge upon public life.

Prewitt's view of scientific literacy requires a different set of assumptions about the selection of content and pedagogy for the biology curriculum. No longer can we assume that the structure of the discipline will provide sufficient guidance; we must, instead, follow Paul Hurd's advice and insist that the context of the learner be the touchstone for the selection of content and teaching strategies.

What is the context of the learner? There are many components, but the essential element for the learner in our society is *change*—rapid and pervasive change in economics, politics, demographics, the home, the workplace, and social mores. Both the rate and direction of change are influenced profoundly by science and technology. The biology curriculum, therefore, must prepare students for a rapidly changing society that is wedded to science and technology. Among the objectives of this curriculum are the following.

- *An understanding of major concepts from a variety of disciplines.* The conceptual boundaries that once separated the major scientific disciplines are fast eroding, and the biology curriculum must acknowledge that one must understand chemistry, physics, and biology to comprehend the impact of science on human affairs and the complexity of the science-related issues that confront us as a collective. Furthermore, the curriculum must inform students that we cannot accommodate rapid change, promote an improved quality of life, or solve science-related social issues with information and expertise from the natural sciences alone. We must introduce students to basic principles from the social and behavioral sciences, so that students understand the critical social and cultural dimensions of our species.

- *An understanding of the history of science as an intellectual and social endeavor.* Contemporary science education is crowded with examples of the history of science, but taken together the examples amount to little more than a poorly articulated chronology of discoveries and inventions. Nowhere in the high-school science curriculum is the student likely to encounter a cohesive picture of the ways in which the intellectual development of the sciences—and of science as an enterprise—shaped history and society and was in turn shaped by them. Science has been and continues to be among the most influential forces in society. It has been responsible for the growth of a rational, empirical view of the natural world that has been instrumental in shaping western society for the last 400 years (Bronowski, 1978).

- *An understanding of the nature of science as an intellectual endeavor.* Science is an attempt by humans to construct rational explanations of the natural world, yet the persistence of widespread belief in astrology, creationism, and other such supernatural nonsense shows that a rational-empirical view of the world is not as pervasive as we might hope. Many American newspapers carry a daily astrology column, while a scant few have even a weekly column on science. The biology curriculum must impress on students that science is a method of rational inquiry into the nature of the universe. The results of this inquiry are always tentative; as Garrett

Hardin (1985) has put it, science is "ineluctably married to doubt." That view is essential to counteract a growing tendency in this country to seek ideologically pure, immutable answers to complex and mercurial problems.

- *An understanding of technology.* Most Americans are likely to encounter science in its technological manifestations and are unlikely to distinguish science from technology. Indeed, it is increasingly difficult even for professional scientists to tell where one ends and the other begins. A recent report prepared by BSCS for the National Center for Improving Science Education (1988) stresses the importance of education *about* technology, not merely *with* technology. The report distinguishes science from technology as follows:

> "**SCIENCE** proposes *explanations* for *observations* about the *natural* world.
>
> "**TECHNOLOGY** proposes *solutions* for *problems of human adaptation* to the environment."

The center's report also provides an overview of basic principles that biology students should understand about technology as a force for change:

> "Technology exists within the context of nature; that is, no technology can contravene biological or physical principles.
>
> "All technologies have unintended consequences.
>
> "Just as proposed explanations about the natural world are tentative and incomplete, proposed technological solutions to problems are incomplete and tentative.
>
> "Because technologies are incomplete and tentative, all technologies carry some risk; a society that is heavily dependent on technology cannot be risk-free."

- *An understanding of the relationships between science and technology and between ethics and public policy.* John Moore (1984) reminds us that science can tell us what we *can* and (more often) *cannot* do, but it is powerless to tell us what we *should* do. The latter question involves values and ethics, where questions of right and wrong—of "oughtness"—dominate the discussions. Students should recognize that ethical analysis is, like scientific analysis, a form of rational inquiry (BSCS, 1988). Unsupported statements and opinions carry no more weight in ethical analysis than they do in science. Ethical analysis is not the sharing of uninformed opinions—what someone once called pluralistic ignorance—but requires instead that we provide well-reasoned arguments for what we ought or ought not to do.

The next step, of course, is public policy, wherein consensus on ethical positions (as well as our imperfect systems can establish it) is expressed as laws and regulations to help to ensure that our ethical vision is translated into actions. Progress in science and technology (genetic engineering and nuclear weaponry, for example) forces us to confront rapid change and raises what were once intellectual abstractions to the level of hard, often painful, reality for individuals, families, and nations. We often must make decisions about new knowledge and technologies that we have barely begun to understand, much less embrace.

- *The ability to use knowledge and solve problems.* If students achieve the foregoing objectives, they will be prepared to use information and the skills of critical inquiry to make decisions and solve problems—for themselves, for their families, for their employers, and for the nation—as informed participants in the democratic process.

The objectives listed above are subsumed by the more global goals of improved quality of life and personal development that are important objectives for general education.

HUMAN ECOLOGY

Which of the many possible conceptual approaches to biology will best help students and teachers to achieve the foregoing objectives? I believe that it is a framework organized on the principles of human ecology. Paul Ehrlich (1985) notes that "human ecology has normally focused on four main areas:

1. the dynamics of human populations;
2. the use of resources by human beings;
3. the impact of human beings on their environment;
4. the complex interactions among 1-3.

Ehrlich proposes human ecology as only part of an introductory undergraduate course in biology. I propose it as a conceptual framework for high-school biology, because it attends to the context of the learner and because it best meets the objectives listed in the preceding section. How might a course in human ecology be structured? What follows are very brief overviews of four hypothetical units of instruction, corresponding to four quarters of the school year. (The assumption that the school year should remain as currently structured is itself open to question, as is the current, year-bound sequence of earth science, biology, chemistry, and physics.)

- *Unit 1—Human Ecology: Population, Resources, and Environment.* This unit helps students to analyze the place of *Homo sapiens* in the biosphere and emphasizes that humans are not exempt from the scientific imperatives that affect all other organisms. Indeed, as Kormondy (1984) points out, human ecology is "not as a kind different from any other kind of ecology, but in degree, the degree to which humans serve in their relationship role" by virtue of their pervasive effect on all other organisms and all other aspects of the biosphere. The unit addresses important concepts that underlie ecological principles, such as reproduction and carrying capacity, the problems inherent in exponential population growth in the presence of finite resources, natural cycles, and the implications of the principles of thermodynamics for the development and use of energy resources (Buchwald, 1984). This unit provides the underlying scientific principles—from chemistry, physics, and biology—for the development of what Hardin (1984) calls "ecolacy . . . the level at which a person achieves a working understanding of the complexity of the world, of the ways in which each quasi-stable state gives way to other quasi-stable states as time passes." The special ways that human beings affect and are affected by those "quasi-stable states" are the focus of this unit; the principles presented are expanded and reinforced in the subsequent units as the principles are applied to specific human problems in an ecological context.

- *Unit 2—Human Behavior: Biological, Psychological, and Cultural Aspects*. This unit explores in detail what is and is not known about the biological and nonbiological determinants of human behavior. Students use data from various subdisciplines of biology, such as genetics and neurobiology, as well as from psychology, sociology, and anthropology, to examine various approaches to the study of human behavior (Konner, 1982). They consider how these different perspectives affect one's view of intelligence, mental illness, biological variation, education, child-rearing, interpersonal relationships, criminality, and the design of human environments. Students consider how knowledge about human behavior might be used to solve social problems.

- *Unit 3—Human Health: Biological, Environmental, and Cultural Aspects*. This unit addresses changing patterns of mortality and morbidity in advanced countries and examines the roles of human biology (especially development and variation), environment, and life style in the determination of personal and community health. The material emphasizes the multifactorial nature of the leading causes of death and disability among adults in developed countries (Sorensen, 1988), as well as the role of risk-taking behavior, accidents, and violence in the health problems of children, adolescents, and young adults (Coates et al., 1982). Students investigate the effect on health of the interactions among genotype, environment, human adaptation, and advances in biotechnology (Holtzman, 1988; BSCS, 1988). Students apply biological principles in cross-cultural comparisons by contrasting health problems in developed countries with those in developing countries, for example, malnutrition and infectious diseases of both humans and livestock. Students examine the ecological relationships that sustain such problems by investigating such concepts as the cultural structure of the population in question, population growth and carrying capacity (Hardin,

1985), and the life cycles of infectious organisms. Students also investigate the potential contributions of such disciplines as genetic engineering and immunology to the resolution of health problems in developed and developing countries alike and consider the problems of introducing such technologies in both settings.

- *Unit 4—Human Adaptation: The Influence of Science and Technology.* Students examine how humans have assumed control of their evolution through the application of science and technology. The material addresses more directly and formally than that in the previous three units the relationships among science, technology, and society and examines how science both derives from and helps to determine societal values. Patrick and Remy (1985) have pointed out that such instruction should help students to "understand the symbiotic relationship of science and technology in order to understand the social context and effects of those distinct and complementary enterprises." Students investigate the growing power and importance of biotechnology, ranging from improvements in agriculture (Office of Technology Assessment, 1988) to the artificial prolongation of life (President's Commission, 1983) and gene therapy using both somatic and germ cells (Office of Technology Assessment, 1984). In each case, students examine both the capabilities and limitations of science and technology and confront the possibility that some problems, such as population growth, may have no technological solutions (Hardin, 1968). Students also explore the growing tendency of technology to influence basic research and, therefore, theory formation (Markle and Robin, 1985; Newman, 1988). The unit addresses basic principles of evolution and adaptation (Cavalli-Sforza, 1983; Bendall, 1983), as well as the special concept of cultural evolution and the transmission of knowledge. Students analyze the role of science and technology, particularly biotechnology, in the creation and resolution of societal problems, as in genetic screening (Holtzman, 1988). The material in this unit stresses the importance of maintaining genetic and cultural diversity (Wilson, 1988) as we apply new technologies and seek resolutions to societal dilemmas. This unit also addresses the various ethical positions that one may assume in considering the relationship of humans to the rest of the biosphere (Morison, 1984; BSCS, 1988) and examines the biological assumptions and consequences of those positions. Students may be asked, for example, to contrast an ethical position that posits the pre-eminence of individual rights with one that favors the rights of society or the state (BSCS, 1988).

The four units proposed here acknowledge the National Science Board's (1983) assertion that "the primary need for the revitalization of biology eduction is perceived to be a conceptual framework that is more in harmony with understanding oneself and which is supportive of the national and global welfare." To that end, the first three units—Human Ecology, Human Behavior, and Human Health—provide a strong basis of scientific concepts and principles, "in terms of the human organism with extension to other life forms." All units help students—in the words of the NSB—to "make responsible use of what they are learning."

INSTRUCTION

Other participants in this conference will address in detail the instructional strategies and technologies appropriate for high-school biology, but the objectives and content proposed herein require some comment about what should be happening in—and outside—the classroom. Students must be *doing* science and using technology, not merely learning *about* science and technology, and they must be *engaged* in discussions of ethics and public policy. One cannot learn skills of critical inquiry passively; one must be involved in constructing one's own knowledge and one's own opinions about issues that matter. The National Assessment of Educational Progress (NAEP) has shown that "eleventh-grade students who reported classroom activities that were challenging and participatory were likely to have higher science proficiency" (Mullis and Jenkins, 1988). Unfortunately, the data show that such instruction is "relatively rare." Improvement of the biology curriculum requires that teachers abandon their traditional role as purveyors of information and become facilitators of learning. It also requires that students collect data of all kinds from outside the classroom.

These suggestions about instruction are not new, but they have not found wide acceptance, partially because they are more time-consuming and difficult than traditional methods (Costenson and Lawson, 1986), partially because students or teachers find no reward for such instruction on standardized tests, and partially because teachers are not trained to teach this way, either formally or through the teaching experienced in their own education (Moore, 1984).

One does not suddenly transform a didactically oriented classroom into an open forum for discussion of the tentativeness of scientific data and the complexities of ethical analysis. One must establish an atmosphere of science as a public inquiry from the first day, and students must expect that they will be challenged continuously in discussions about hypothesis formation, the structure of investigations, interpretation of data, and the implications of one's results or values.

The issues raised in a course whose framework is human ecology will sometimes be controversial. Teachers, administrators, publishers, and parents must get used to that fact, because scientific and technological progress induces controversy as a matter of course. From the evolutionary thread that must permeate any approach to biology to discussions of genetic screening and selective abortion, students will confront complex and contentious issues. Controversy should not be the focus of the course, but neither should we avoid controversy if it surrounds some topics. Teachers must be trained to handle controversy in the classroom and to lead activities and discussions that help students to examine all sides of a given issue.

THE LOYAL OPPOSITION

Because the amount of opposition to change is generally directly proportional to the degree of change proposed, there will be considerable opposition to my proposed restructuring of the high-school biology curriculum. Substantial inertia in the educational system militates against change. For example, more than 30% of teachers indicate that they are satisfied with current biology textbooks (Weiss, 1987), notwithstanding that those textbooks receive extremely poor grades from scientists and science educators (Johnston, 1988; McInerney, 1986). Publishers, who must agree to change if there is to be any improvement in the curriculum, have no incentive to change and, in fact, are rewarded if they do not change (Apple, 1985; Tyson-Bernstein, 1988).

There will be at least three major objections to my proposal to make human ecology the focus of the high-school biology program:

- *"It is not science."* Some will criticize the emphasis on human ecology because students will spend some of their time on issues of ethics and public policy as they consider how to manage problems related to science and technology. Students must have a substantive content base, because one cannot consider matters of bioethics and public policy without a sound understanding of the science (BSCS, 1988). The content we choose for the biology curriculum, however, should promote the skills of rational inquiry that will stand students in good stead beyond an hour-long examination that tests trivial knowledge derived from trivial teaching and trivial textbooks. I reiterate that the content and pedagogy should reflect the current and future context of the learner—change—and the requirements of scientific literacy outlined by Prewitt (1983). Should we have rote recitation of the stages of mitosis, or should we have a problem-oriented look at the environmental factors that damage genetic material, the progress we are making in detection and treatment of genetic disorders (White and Caskey, 1988), and the ethical and policy implications thereof (Holtzman, 1988)?

- *"The approach is anthropocentric; what happens to the rest of the organisms we teach about?"* This criticism fails on two counts. First, it presumes that human ecology does not encompass other organisms. The third component in Ehrlich's definition of human ecology is "the impact of human beings on their environment." This, of course, assumes that we know what is in the environment and that we recognize that the principles of chemistry, physics, and biology that apply to humans apply to other organisms as well. Second, criticism on grounds of anthropocentrism presumes—as do most textbooks—that students must be intimately acquainted with the details of all major taxonomic groups. What results is a forced march through the phyla, rather than a problem-oriented look at diversity, evolutionary and ecological relationships, and the danger that inures to us all by virtue of the ceaseless assault on the environment (Wilson, 1988; May, 1988; Partridge and Harvey, 1988; Lande, 1988).

- *"It is not rigorous enough."* If there is not enough "content," by which most people mean "facts," some will assume that the program is appropriate only for "academically unsuccessful learners" and clearly not for those who are college-bound, especially if those students are to study science. Bybee (1984), however, in emphasizing the importance of human ecology in biology education, stated:

> Courses, units, or lessons with an emphasis on human ecology should be required of *all* students. Neither are these advanced placement, accelerated, or second-level programs, nor are these programs exclusively for slow learners, low track, or vocational students.

We should beware a false sense of rigor, such as that implied by the "back to basics" movement. This conceptualization of rigor is limited intellectually, because it demands nothing more than low-level skills, and limited educationally, because it does not prepare students for life in a complex, technological society. Life in contemporary society requires an intellectual rigor whose hallmarks are critical thinking and problem-solving—skills that will stand students in good stead in the workplace, in the voting booth, and in the home.

The 1986 NAEP (Mullis and Jenkins, 1988) assessment shows that most of the improvement in science performance—where there was any improvement at all—came in the areas of "lower-level skills and basic science knowledge." To be sure, those results are partially a function of the ease of assessment of such skills. But the results likely reflect as well the emphasis on such skills in textbooks and therefore in the classroom. In contrast, the 1983 recommendations of the National Science Board (1983) called for "new science and technology courses that are designed to meet new educational goals . . . [and] that incorporate appropriate scientific and technological knowledge and are oriented toward practical issues."

The NAEP report confirms that "what has traditionally been taught in science may be neither sufficient nor appropriate for the demands of the future, necessitating reforms that go beyond increasing students' exposure to science and that center on implementing new goals for improving curriculum and instruction."

AN INTERNATIONAL PERSPECTIVE

There is an underlying theme of international competitiveness in our present approach to the restructuring of science education, and in some ways that has been helpful. Concern about flagging American performance has served as a vehicle for bringing education to the attention of policymakers and the public, and attempts at improvement likely would have found scant political support had they not been framed in the need to sustain economic and military advantage. Both economic and military issues, of course, ultimately have their roots in resource issues (Ehrlich, 1985; Hardin, 1985), and a focus on competition in the international arena presumes that there will be something of perpetual value that merits such competition. Unless we act to reverse the trend of "living on our capital" of natural resources (Ehrlich, 1985), however, that assumption is by no means sound.

I labor the obvious to state that the basic principles inherent in a course in human ecology are unencumbered by national boundaries. Indeed, I think it imperative that we broaden our focus to involve representatives of as many nations as possible in the conceptualization of such a course. A recent meeting of science educators from 40 countries confirmed the universal need for a change in the content and methods of science education. Although the problems of developing countries differ from those of the developed world, science educators around the world recognize the impact of science and technology on rapid and continuous change, and they feel that their citizens must be prepared to manage that change. The details of the curriculum will differ from country to country, but I believe that we can reach rapid and easy agreement on the principles that citizens of all nations must understand if there is to be anything left on the planet worth competing for.

EVOLUTION OR REVOLUTION

We Americans proudly proclaim that we do not have a national curriculum and delight in the decentralization of curriculum decisions such that each state is free to establish its own guidelines and each district in a state is free to structure its courses to meet those guidelines. The control of the curriculum by a few major textbooks (Weiss, 1987) and the similarity of those books (Gould, 1988; McInerney, 1986) put the lie to that assertion, particularly given the extent to which the textbook determines course content (Muther, 1985; Tyson-Bernstein, 1988). The fact is that we do have a national curriculum in biology, and it is failing and in need of wholesale change.

Change is traumatic and difficult, particularly in the educational system, which is beset by inertia and by the tendency to protect vested interests. Even many of those who acknowledge that change is necessary suggest that gradual, incremental change is the best approach to restructuring the biology curriculum. This is a gradualistic evolutionary model that assumes the slow, steady accumulation of variation and low levels of speciation. The arguments that support this approach include the need to allow the system to respond slowly and deliberately to selection pressure, testing out, as it were, each new curricular phenotype in the environmental crucible of the classroom. That would be a reasonable approach if the rate of environmental change were low, the direction of change were not substantially at odds with the current environment, and there were likely to be enough variation in the population of curricular approaches to allow legitimate selection.

The rate and direction of societal change induced by science and technology argue, in fact, for punctuated equilibrium—relatively rapid development of new species of curriculum in response to substantive environmental pressure. We do not need any more evidence than that already accumulated to convince us that our present approach to education—the biology curriculum included, perhaps most especially—is not meeting the needs of learners. The 1986 NAEP assessment (Mullis and Jenkins, 1988) states that "radical change" is required in the nation's schools if today's elementary-school and middle-school students are to reverse the poor performance of today's high-school students.

Some say that relatively rapid change is not possible, but they rarely tell us why. The naysaying generally amounts not to cogent arguments, but to what Richard Dawkins (1986) calls "affirmations of incredulity." These opinions have no real foundation in fact and provide no insights into improvement of the situation.

We now have an opportunity to promote revolution by developing the first step on the road to an integrated science that reflects more accurately the status of modern science and that meets the needs of learners. There is no question but that the revolution will be costly: new books and new technology; new assumptions about teaching and the training required to bring teachers and administrators up to speed; education of parents, who will see little in science that they recognize from their own courses. But we can hardly afford the alternative, which is stasis.

REFERENCES

Apple, M. W. 1985. Making knowledge legitimate: Power, profit, and the textbook, pp. 73-89. In Current Thoughts on Curriculum. Alexandria, Va.: Association for Supervision and Curriculum Development.

Bendall, D. S., Ed. 1983. Evolution from Molecules to Men. Cambridge, England: Cambridge University Press.

Bronowski, J. 1978. Magic, Science, and Civilization. New York: Columbia University Press.

BSCS (Biological Sciences Curriculum Study). 1988. Advances in Genetic Technology. Lexington, Mass.: D. C. Heath.

Buchwald, C. E. 1984. Human ecology: A first lesson. Amer. Biol. Teach. 46:330-333.

Bybee, R. W. 1984. Human Ecology: A Perspective for Biology Education. Reston, Va.: National Association of Biology Teachers.

Cavalli-Sforza, L. L. 1983. The Genetics of Human Races. Burlington, N.C.: Carolina Biological Supply Company.

Coates, T. J., A. C. Petersen, and C. Perry, Eds. 1982. Promoting Adolescent Health. New York: Academic Press.

Costenson, K., and A. E. Lawson. 1986. Why isn't inquiry used in more classrooms? Amer. Biol. Teach. 48:150-158.

Dawkins, R. 1986. The Blind Watchmaker. New York: W. W. Norton and Company.

Ehrlich, P. R. 1985. Human ecology for an introductory biology course: An overview. Amer. Zool. 25:379-394.

Gould, S. J. 1988. The heart of terminology. Nat. Hist. 97(2):24.

Hardin, G. 1968. The tragedy of the commons. Science 162:1243-1248.

Hardin, G. 1984. An Ecolate View of the Human Predicament. Washington, D.C.: Environmental Fund.

Hardin, G. 1985. Human ecology: Subversive and conservative. Amer. Zool. 25:469-476.

Holtzman, N. A. 1988. Recombinant DNA technology, genetic tests, and public policy. Amer. J. Hum. Gen. 42:624-632.

Johnston, K., Ed. 1988. Science textbook update. Sci. Books Films 28:199.

Konner, M. 1982. The Tangled Wing: Biological Constraints on the Human Spirit. New York: Harper and Row.

Kormondy, E. 1984. Human ecology: An introduction for biology teachers. Amer. Biol. Teach. 46:325-329.

Lande, R. 1988. Genetics and demography in biological conservation. Science 24:1455-1460.

Markle, G. E., and S. R. Robin. 1985. Biotechnology and the social reconstruction of molecular biology. BioScience 35:220.

May, R. M. 1988. How many species are there on earth? Science 24:1441-1449.

McInerney, J. D. 1986. Biology textbooks: Whose business? Amer. Biol. Teach. 48:396-400.

McInerney, J. D. 1987. Curriculum development at the Biological Sciences Curriculum Study. Educ. Leader. 44(4):24.

Miller, J. D., R. W. Sucher, and A. M. Voelker. 1980. Citizenship in an Age of Science. New York: Pergamon Press.

Moore, J. A. 1984. Science as a way of knowing: Evolutionary biology. Amer. Zool. 24:467-534.

Morison, R. S. 1984. The biological limits on autonomy. Hastings Center Rep. 14(5):43-49.

Mullis, I. V. S., and L. B. Jenkins. 1988. The Science Report Card: Elements of Risk and Recovery. Princeton, N.J.: Educational Testing Service.

Muther, C. 1985. What every textbook evaluator should know. Educ. Leader. 42(7):4.

National Center for Improving Science Education. 1988. Science and Technology Education for the Elementary Years: Curriculum and Instructional Frameworks. Andover, Mass.: The Network.

National Commission on Excellence in Education. 1983. A Nation at Risk. Washington, D.C.: U.S. Government Printing Office.

National Science Board (Commission on Precollege Education in Mathematics, Science, and Technology). 1983. Educating Americans for the 21st Century. Washington, D.C.: National Science Foundation.

Newman, S. A. 1988. Idealist biology. Perspect. Biol. Med. 31:353-368.

Office of Technology Assessment. 1984. Human Gene Therapy—A Background Paper. Washington, D.C.: U.S. Government Printing Office.

Office of Technology Assessment. 1988. New Developments in Biotechnology—Field Testing Engineered Organisms: Genetic and Ecological Issues. Washington, D.C.: U.S. Government Printing Office.

Partridge, L., and P. H. Harvey. 1988. The ecological context of life history evolution. Science 24:1449-1455.

Patrick, J. J., and R. C. Remy. 1985. Connecting Science, Technology, and Society in the Education of Citizens. Boulder, Colo.: Social Sciences Education Consortium.

President's Commission (for the Study of Ethical Problems in Medicine and Biomedical and Behavioral Research). 1983. Genetic Screening and Counseling. Washington, D.C.: U.S. Government Printing Office.

Prewitt, K. 1983. Scientific illiteracy and democratic theory. Daedalus 112:49-64.

Rutherford, F. J., and A. Ahlgren. 1988. Rethinking the science curriculum, pp. 75-90. In R. S. Brandt, Ed. Content of the Curriculum. Alexandria, Va.: Association for Supervision and Curriculum Development.

Sorensen, T. I. 1988. Genetic and environmental influences on premature death in adoptees. New Engl. J. Med. 318:727-732.

Tyson-Bernstein, H. 1988. A conspiracy of good intentions: America's textbook fiasco. Washington, D.C.: Council on Basic Education.

Weiss, I. R. 1987. Report of the 1985-86 National Survey of Science and Mathematics Education. Research Triangle Park, N.C.: Research Triangle Institute.

White, R., and T. Caskey. 1988. The human as an experimental system in molecular genetics. Science 240:1483.

Wilson, E. O., Ed. 1988. Biodiversity. Washington, D.C.: National Academy Press.

15
Developing a Synthesis Between Seventh-Grade Life Science and Tenth-Grade Biology

WAYNE A. MOYER

My thesis is simple. Life science—commonly taught in the seventh grade—is nothing but a watered-down version of tenth-grade biology. It does not have to be, but most teachers approach the course from that viewpoint. Furthermore, the available textbooks patronize naive students by oversimplifying complex ideas and feeding them conclusions from which all intellectual juice has been squeezed. The result is a course heavy in vocabulary and brute memorization—justified with the argument that "you will need to know these terms when you take biology." Presumably, it will be *real* biology, for which "life science" has been but an introduction. In fact, it is likely to be just another survey of traditional biology.

To help convince you that this pessimistic picture accurately describes the current state of affairs, let me share with you instructional objectives for life science and biology stated in the Montgomery County (Maryland) public-school program of studies. Table 1 compares the instructional objectives for the topics of cells, levels of organization, reproduction, and taxonomy. The similarity is obvious. When we compare all the topics covered in the two courses, we find remarkable overlap. Life science and

Wayne A. Moyer received a Ph.D. in developmental biology in 1974 from Princeton University. He is coordinator of secondary science, Montgomery County (Maryland) Public Schools, and was the director of the Math/Science Clearinghouse, PRISM, in Philadelphia in 1985-1987; science director of People for the American Way in 1983-1985; and executive director of the National Association of Biology Teachers in 1979-1983. He had been with the Seton Hall University Biology Department in 1977-1979.

TABLE 1 Comparison of Instructional Objectives for Life Science and Biology, Montgomery County Public Schools

Topic	Objectives in Life Science	Objectives in Biology
Cells	Compare animal and plant cells; state functions of: nucleus, cell membrane, cytoplasm, cell wall, chloroplast, and vacuole	Investigate general structures, functions, biochemistry, and diversity of cells
Organization	Arrange biological level of cellular organization from least to most complex	Describe various levels of organization in living systems
Reproduction	Describe major differences between sexual and asexual reproduction	Investigate perpetuation of species through sexual and asexual reproduction
Taxonomy	Match organisms with their phyla; use dichotomous keys to name organisms	Apply methods of taxonomy to classification of major groups of organisms

biology both cover cells, organization, reproduction, human anatomy, genetics, taxonomy, germ theory, botany, behavior, and ecology. Drug abuse and nutrition are covered in life science, but not in biology. The only topic in tenth-grade biology *not* covered in life science is evolution. Evolutionary theory is also absent from several popular life-science textbooks.

When we compare textbooks written for life science and biology, we find the same underlying assumption: life science is watered-down biology. The following extracts show what two textbooks have to say about aspects of cell theory.

From *Life Science* (Ramsey et al., 1986, pp. 44-45):

> *Basic Cell Structure*
>
> Cells are made of protoplasm and its products. Cells are not all the same size and shape. Many cells have special structures that have special purposes. But all cells are similar in some respects.
>
> Surrounding the cells is a covering called the *cell membrane*. See Fig. 2-12. It controls what materials enter or leave the cell. Most of the cell is made of a type of protoplasm called *cytoplasm*. Many of the cell's activities are carried on in the cytoplasm. Near the center of the cell is a structure called the *nucleus*. The nucleus is the "control center" that directs all the cell's activities. It is surrounded by a *nuclear membrane*. Inside the nucleus is a type of protoplasm called *nucleoplasm*.

From *Modern Biology* (Otto and Towle, 1985, pp. 56-57):

> *4.4 Parts of a Cell*
>
> Cells are very complex and vary in size and shape. Each cell is surrounded by a cell membrane or a plasma membrane. This flexible membrane separates the inside of the cell from its surroundings. In some cells such as the ameba, this membrane is very flexible and the ameba may change its shape. Another characteristic of cells, except those of bacteria and the blue-green bacteria, is that they each contain a large oval or spherical body. This is the nucleus. The nucleus is the control center for all cell activity. Look at the cells in figure 4.1 and observe the cytoplasm. The cytoplasm consists of the cell material between the nucleus and the cell membrane. Small structures in this area are suspended in the cytoplasm.

Note the archaic word "protoplasm" used freely in *Life Science.* "Control center" is used as a metaphor for nuclear function in both books. Note also the gratuitous introduction of the useless term "nucleoplasm." Finally, note the stilted prose, so typical of textbooks written to meet the requirements of a reading formula.

Turning to the *Modern Biology* introductory paragraph on cells, one can at least be pleased that "protoplasm" is gone. Yet the authors, in attempting to present a brief summary of cell structure, oversimplify and thereby create erroneous images in the reader's mind. For example, the ability of an ameba to change its shape is attributed to the plasma membrane, rather than to internal structures. Furthermore, the complexity of intracellular architecture is blurred by referring to "small structures . . . suspended in the cytoplasm."

In summary, even this brief analysis suggests that life-science textbooks tend to be out of date, present an oversimplified view of biology, and mimic the structure—if not the wording—of biology textbooks. We also observe stilted prose that is difficult to understand.

We all know that curriculum guides and textbooks do not necessarily reflect the day-to-day activities of a science classroom. Weak guides and flawed textbooks can be interpreted by imaginative teachers to produce exciting courses. Here, then, are a few examples of activities I have observed in life-science classrooms.

- Prepare a report on a disease of the student's choice.
- Dissect a frog, beginning with external observations and progressing on succeeding days to internal organs.
- Conduct an environmental hearing before a jury of students, with presentations on both sides of an issue.
- View a filmstrip on pollution in Chesapeake Bay.
- Write definitions of anatomical terms on a worksheet.
- Dissect a flower and name the parts.
- Take a practical laboratory quiz on frog anatomy.
- Identify an unknown phylum by means of "yes" and "no" questions.

- Dissect an earthworm and identify the parts.

As you can see, this is a mixed bag of activities, but probably quite typical of those found across the country. They range in difficulty from filling in blanks on a worksheet to developing arguments for and against action on an environmental issue, from emphasis on memorizing vocabulary to developing higher-order intellectual skills. However, except for a few activities, all would be equally appropriate for a biology class. In fact, most of the activities *are* repeated in the tenth grade. No wonder students complain that science is boring! How many times should a student dissect a frog or identify an earthworm as a member of the phylum Annelida? The thought of hundreds of students dissecting hundreds of preserved frogs in the belief that they are studying the science of life is troubling. William Mayer calls this necrology, instead of biology.

Last summer, seven experienced teachers met for 2 weeks to consider the following question: What should every graduate of Montgomery County public schools know about science and technology, and when should it be taught? In effect, this meant taking a close look at the science taught in grades 7 through 10, which constitutes the common core of scientific knowledge acquired by every student. Their mandate was to view these science courses as a single system and to present a plan for future curriculum development.

Their primary reference was a draft copy of the Project 2061 Phase I Report, kindly provided by James Rutherford of the American Association for the Advancement of Science. Titled *Science for All Americans*, it contains the reports of several task forces that have been deliberating since the 1986 apparition of Halley's Comet (American Association for the Advancement of Science, 1989). The project director plans to publish the report this year, long before the comet's next apparition in 2061.

The work group began by developing 15 statements of philosophy or objectives, which served to define the type of science instruction every student should receive (Table 2). In summary, they envision an activity-centered curriculum that draws content toward it as required, rather than a content-centered curriculum with activities traditionally hung on the content framework like decorations on a Christmas tree. This simple reversal of the traditional order should have profound effects on science instruction. Textbooks will become references—along with computer-managed databases, video disks, and periodicals—rather than being the curriculum itself. Teachers will become facilitators and co-investigators, rather than fonts of knowledge. And classrooms will look outward to the world, rather than inward to vocabulary lists.

In practice, the work group proposed that each instructional unit include a unifying activity, or focus, that would serve to tie the content

TABLE 2 Objectives--Summary Statements

1. Learning science should be related to the student's everyday experiences.
2. Learning science should be an activity-based, stimulating process.
3. Students should be given every opportunity to attain success and develop a positive attitude toward science.
4. Students should observe and participate in activities that encourage creativity.
5. Students should be encouraged to develop a healthy skepticism.
6. Students should have hands-on experiences that relate science and technology.
7. Science instruction should reflect the interdisciplinary nature of learning.
8. Students should have multiple opportunities to test hypotheses by collecting, describing, and interpreting data.
9. Students should perceive science as a cooperative effort.
10. Students should be provided ample time to explore, observe, and assess the science processes.
11. Every student should be challenged with problems that require higher-order thinking skills to reach solutions.
12. Students should develop a knowledge base that supports the structure of science disciplines.
13. Students should be prepared to deal responsibly with societal issues related to science and technology.
14. Students should have a variety of science experiences aimed at providing a basis for exploring and planning careers.
15. Science instruction should make use of appropriate resources in the community.

to everyday experience, require application of higher-order thinking skills, and involve societal issues related to science and technology—in short, an overarching activity that would implement the objectives. For life science, such an activity might be development of a model spacecraft that would support human life for an extended period in space. The work-group participants suggested that the project be organized as a cooperative effort within a class, with small groups considering various aspects, such as waste disposal, recycling, and environmental requirements. The activity would be included in the unit on human physiology.

The overall emphasis of the "new" life-science course would be the human animal, and the primary experimental organism would be the student. This plays straight to the interests of seventh-graders: Who am I? What am I becoming? The proposed content closely follows the Project 2061 recommendations, with the addition of units on plants and agriculture. The course is whimsically called, "Humans and Beans" (Table 3). The year concludes with a study of problems related to the human presence on Earth. In the words of a work-group participant, "this gives students an opportunity to focus on issues of science, technology and society, and to examine their personal roles in shaping the world of the future."

TABLE 3 "Humans and Beans"

	Additional Plant Topics
Plants:	
Structure and function	
Maintenance	
Origins:	
Human history	Domestication of plants
Genomes and gene pools	
Variation	
Characterization and classification	
Evolution	
Life cycle:	
Reproduction	
Differentiation	
Development	Seed-plant development
Maturation	Hormone influence and auxins
Aging	
Functions:	
Homeostasis	
Organ systems	
Feedback mechanisms	
Energy requirements	
Nutrient requirements	Food plants
Learning process:	
Skills	
Behavior	
Physical health:	
Definitions	
Maintenance and homeostasis	
Disorders, symptoms, and treatment	Plant diseases--rusts and blights
Germ theory of disease	
Mental health:	
Cultural	
Social	
Coping mechanisms	
Stress and prolonged disturbance	Plant maintenance--wilt, turgor, and life expectancy
Treatment	
Death and dying	
Human presence:	
Population	
Resources	Forestry and food supply
Survival	

The task of our work group was to consider the science taught in grades 7 through 10 as a single system. Thus, what is taught in grade 7 should not be retaught in grade 10—reviewed, yes, preferably through independent reading and computerized tutoring programs, but not retaught. Each course, life science and biology, must therefore stand alone—each tub on its own bottom.

Working within these limits, the work group decided that tenth-grade biology ought to be a fairly sophisticated course that sets forth the conceptual framework of a major field of science. The proposed content is not radically different from that of a traditional biology course (Table 4), although some knowledge of chemistry, physics, and earth science is assumed.

Notice that this proposed biology curriculum reverses the normal order of topics. Rather than starting with molecules and cells, it begins with a look at organisms and ecosystems. In this regard, it resembles *Biological Science: An Ecological Approach,* the Biological Science Curriculum Study green version (Biological Sciences Curriculum Study, 1987). Only in the second semester does the student get to cells, genetics, and energy. A final review of evolutionary theory becomes the culminating unit and serves to unite all of biology under one explanatory theory.

This is but the first step in our attempt to unite life science and biology

TABLE 4 The Living Environment (Proposed Tenth-Grade Biology Curriculum)

First Semester	Second Semester
Interdependence:	Cells:
Interactions and interrelationships	History of cellular biology and technology
Environments	Cell structure and function
Population density	Homeostasis and cellular control
Equilibrium	Genetics:
Characterization:	Mendelian genetics
Classification	Human genetics
Speciation	Modes of reproduction
Categories	Molecular genetics
Products	Flow of matter and energy:
Multicellular organisms:	Energy sources
Multicellular systems	Energy pathways
Growth and development	Cycles (biogeochemical)
Differentiation	Conservation
Reproduction and life cycles	Pollution
Behavior:	Evolution:
Kinds of behavior	Definition
Hierarchies	Factual considerations
	Gene frequencies
	Extinctions
	Origin-of-life theories

into a rational curriculum for all students. We have not yet considered how and where to introduce concepts from the physical sciences. Nor have we agreed that all topics now listed should actually be covered. However, we strongly agree that less is better, provided that what is covered is truly learned by students. If the philosophy statements are translated into practice, we believe that this will be the case.

REFERENCES

American Association for the Advancement of Science. 1989. Science for All Americans. Washington, D.C.: American Association for the Advancement of Science.

Biological Sciences Curriculum Study. 1987. Biological Science: An Ecological Approach. 6th ed. Dubuque, Iowa: Kendall/Hunt Publishing Co.

Otto, J. H., and A. Towle. 1985. Modern Biology. New York: Holt, Rinehart and Winston.

Ramsey, W. L., L. A. Gabriel, J. F. McGuirk, C. R. Phillips, and F. M. Watenpaugh. 1986. Life Science. New York: Holt, Rinehart and Winston.

16

Biology Education: Asking the Right Questions

FRANCES S. VANDERVOORT

The title of my presentation is "Biology Education: Asking the Right Questions." What *are* the right questions for biology educators to ask? I offer the following:

- How much biology should be taught?
- What can we learn from the past?
- What kind of biology should be taught?
- What is the social importance of biology education?

HOW MUCH BIOLOGY SHOULD BE TAUGHT?

A few years ago, I attended a lecture by Victor Weisskopf (1984), the distinguished physicist from the Massachusetts Institute of Technology. In this lecture, which focused on the critical state of science education, he described how, as an 8-year-old child in Vienna, he was walking with his father in the forest. He saw a bird and said, "Father, what is that bird's name?" His father chided him. "Do not ask that question, my son," he said. "The essential thing about that bird is *not* its name, but that it

---

Frances S. Vandervoort received a B.S. and M.S. in zoology in 1957 and 1965 from the University of Chicago. She was an instructor at the University of Illinois, Chicago, in 1965-1967 and at Chicago State University in 1969 and 1974. She has been a teacher of biology and physical science in the Chicago public schools since 1975. She was the recipient of the Illinois Governor's Master Teacher Award in 1984 and was an Illinois finalist for the Presidential Award in 1987.

flies, that it has wings, that it lives!" In other words, do not trivialize this wonderful animal by being concerned only about its name.

Weisskopf offered these words of advice: "Begin teaching," he said, "by asking questions." "Ask questions," he said, "but don't give answers. Teachers cannot give definite answers to questions and students must learn not to expect them to. Students learn poorly if teachers attempt to press information into their brains."

Weisskopf went on to say that, when students ask him how much of the subject he expects to "cover" during a course, he answers that he never attempts to "cover" a subject. Instead, he promises to "uncover" parts of it. Students must learn that science *is*, not that it *covers* something. Weisskopf encouraged all teachers to "foster the joy of insight." For this, he said, "the question is the key. We must never lose sight of the social significance of this."

What is the origin of this idea that teachers should regard a young person's brain as an empty vessel to be furnished with facts, rather than a uniquely specialized organ to be carefully nurtured and trained? One problem is the enormous productivity of the scientific community. Today, high-school biology textbooks average 450 pages in length and contain as many as 2,400 new terms, far more than a first-year foreign-language course. Publishers feel compelled to provide students with information about all the latest scientific discoveries. What, then, do they dare leave out from previous editions to make room for the new material? The answer is usually—nothing.

The other day, I happened across an article in *U. S. News and World Report* entitled "Drowning in a Sea of Knowledge" (Allman, 1988). The article described the flood of scientific papers published in the thousands of scientific journals now on library shelves. In this article, one scientist commented that, "If 80 percent of the papers weren't written, the progress of science wouldn't be affected at all." First-year biology students must indeed feel as if they are drowning when confronted with the deluge of detail in so many of today's biology textbooks. What would be the effect on biology education if publishers decided to reduce by 80% the additions they make to new textbooks? I am convinced that teachers, students, and publishers would all benefit from such a step!

For some reason, I seem unable to "cover" as much material as other biology teachers in my department. I sometimes regret not finding time to teach more physiology or anatomy. I enjoy these subjects and think my students would enjoy them as well. I like to think I make up for these omissions by taking time for inquiry-based activities. Some of these are the "Invitations to Inquiry" from the *Biology Teachers' Handbook* (Mayer, 1970), and some I have prepared myself. If you are not familiar with them, these open-ended discussion sets provide excellent opportunities for students to practice scientific thinking. Each inquiry takes at least an entire class period to complete. Students find them very satisfying to do.

Another reason I don't "cover" so much material as other teachers may be that the students themselves try to slow me down. This is not so much because they are lazy (they are not) or because they are overloaded with homework in other subjects. Instead, it seems that they develop a genuine interest in what we are doing and simply don't want to leave it.

I often feel myself rushing on, faster than I wish, knowing full well that much of what I teach will be forgotten before my students graduate from high school. Why do teachers do it? Why do I do it? Why is the emphasis in biology still on the amount of material "covered," rather than on how much the students learn about the processes of science?

One argument for teaching a content-oriented course is that this "prepares" students for the next level of science offered in the school. Unfortunately, overemphasis on detail too often ensures that the student will never again have interest in taking other science courses. In fact, in their view, excessive detail can actually *trivialize* science. How can they learn of the importance of a crayfish to a wetland ecosystem when all they are made to do is remember the number and kinds of legs a crayfish has? Inevitably, they come to regard biology and other areas of science as irrelevant to their lives or too complicated to understand, even if they suspect that it *is* relevant.

How has the state of biology education progressed to the point where textbooks are so thick that students can hardly carry them home? Why have laboratory exercises degenerated to where they are little more than cookbookery for which the end result is obvious to students, even before they walk into the laboratory? How can students learn the processes of scientific investigation when they are served whole meals of scientific facts, rather than being invited into the kitchen, where genuine discoveries are made? To gain a view into this, let us take a brief look at the history of biology education in the United States.

WHAT CAN WE LEARN FROM THE PAST?

Until the 1850s, biology, then termed "natural philosophy" or "natural theology," was studied in this country and in Europe mainly by scholars and theologians who sought to understand better the marvels of God's perfect world. Biology was not taught as a separate course in high schools in the United States until the turn of the century. Then, zoology, botany, and physiology were combined to provide the single, more comprehensive course we now call "biology." Biology soon became the science course of choice of most high-school students. Early biology courses included, among more conventional topics, discussion of human welfare, health, and sanitation. During the Great Depression, biology courses responded to the times by offering consumer education, social welfare, and agricultural science.

Until the late 1950s, high-school biology could best be defined as descriptive, rather than experimental. The role of the teacher was primarily that of transmitter of knowledge. Students approached the study of living things systematically by noting, observing, and describing the external and internal characteristics of a "typical" representative of the phylogenetic group under consideration. The high point of the year came in spring, when students were given a frog to dissect. Laboratory experiments were designed to verify existing knowledge. In short, students learned about the *products* of scientific research, but very little about the scientific *process*.

I have a strange sense of déjà vu as I write this, because these statements about pre-Sputnik science are almost identical with what is being said about biology education today. Did we learn anything from our experiences of the first half-century? Or have we come full circle?

In the decade after World War II, science educators began to recognize that science education must be freed from the intellectual strait jacket in which it had been so long confined. Sputnik was the ultimate catalyst: the federal government began giving top priority to the development of programs in science education that would "put us ahead of the Russians." The Biological Sciences Curriculum Study (BSCS) was merely one facet of a vast effort to upgrade the status of science education in the United States. New laboratory materials and procedures were developed. Workshops funded by the National Science Foundation prepared teachers for using the new materials. Educators began using new learning theories and techniques of investigation. By 1970, most of the nation's schools were using BSCS materials. Underlying this massive effort was the conviction that science must be taught as a process of investigation and inquiry, rather than as accretion of rigid facts and rules.

Public support for science education began to diminish in the early 1970s. Reasons for this are complex, but include, among other factors, the rejection of science and technology because of their close association with the war in Vietnam. Also, BSCS programs had opened Pandora's box by placing so much emphasis on evolution. Christian fundamentalist groups rebelled by bringing pressure on school boards that used the new materials. Sales of BSCS materials dropped precipitously (Hurd, 1980). As public interest in science waned, financial support lessened, until, in the early 1980s, alarms again were sounded. Once again science education had reached a crisis stage. And again we hear criticism that science courses are too rigid, too content-oriented, too inclined toward passive inculcation of students.

WHAT KIND OF BIOLOGY SHOULD BE TAUGHT?

This year, I share my classroom with a biology teacher new to the system. Recently graduated from college, she is attractive and enthusiastic, and I am convinced that the future bodes well for her. The other day, she asked me whether I knew where the scalpels, forceps, scissors, and other dissecting equipment were kept. I professed to not being certain where these items were, because, as I explained, I seldom ask my student to dissect. She could barely contain her astonishment. I responded, to her surprise, by commenting that I have found many ways to teach biology without using preserved specimens. Some educators refer to the excessive use of preserved specimens as "morgue science." I wouldn't go that far, but it is with a measure of satisfaction that I note that science teachers' journals are encouraging teachers to abandon tradition—when a live animal is available for use, *don't dissect* (Berman, 1984)!

If we grant that teachers cannot effectively teach all the material in a standard biology textbook, how can we decide what should be taught? If we agree to de-emphasize, say, anatomical details, chemical formulas, reproductive cycles, and the like, what *should* be taught?

There is no simple answer to this. All teachers have pet topics to teach, and most have some that they would prefer to avoid. There is, however, a backbone of biological thought based on the three great theories of biology: cell theory, gene theory, and evolutionary theory. These theories must be the foundation of all biology education. As I describe these theories to my students, I like to compare them with the three legs of a great tripod supporting all of biology. These three struts are necessary for understanding life on earth; remove any one of them and the whole structure of biology crumbles. They are—*all three*—essential for the teaching of biology.

It is, of course, essential that students understand what is meant by the term "theory." Textbooks don't always help in this matter. "Theory" is a sophisticated concept, and too often textbooks convey the impression that a theory is little more than a casual conjecture. "It's just a theory," one might hear in a soap opera, that Elaine has fed strychnine to Jennifer because she suspected that Jennifer was seeing Robert, her own flame, on the sly. Also, it doesn't help that in 1980, presidential candidate Ronald Reagan stated before a sympathetic audience in Texas that "evolution is just a theory, only one of several theories about the origin of life."

One of the more commonly used high-school biology textbooks asserts that "there are many theories, or ideas, as to how life began on earth, including . . . the Greek myths and some American Indian legends." This book also labels as a "theory" the hypothesis (and it *is* a hypothesis) that life came to earth from elsewhere in the universe. Finally, the book invites students to conduct a poll of 10 people to determine their *theories* about how life originated on earth. Here we have a book, from a reputable publisher, expecting us to ask our students to determine by consensus what is and is not science!

Good texts and good teachers will provide students with a framework for developing an understanding of the nature of scientific theory. As we know, the three traits of all scientific theories are that they are predictive, testable, and tentative. Students are capable of understanding these concepts, and it is satisfying to help them to do so.

In addition to the three main theories of biology, a particularly useful theory for high-school biology teaching is the cell symbiosis theory, promoted most notably by Lynn Margulis of Boston University. In the early 1970s, a former professor of mine at the University of Chicago, who had also taught Lynn, handed me a book—in fact the first book she had written. In it, she first advanced the evidence that she had gathered for the theory of the origin of eukaryotic cells from the symbiotic combining of various types of prokaryotes. Today, this theory is included in many high-school biology texts and is widely accepted by the scientific community. When she first began publicizing her work in the mid-1960s, her ideas were regarded with benign amusement, if not with scorn. You know what Thomas Huxley said: "It is the customary fate of new truths to begin as heresies and end as superstitions" (cited in Oxford Dictionary of Quotations, 1980). I doubt that the cell symbiosis theory will end as superstition, but early on it certainly was regarded as somewhat heretical. We now know the theory for the excellent science it represents; our students should be familiar with it as well.

I must also mention the latest theory with which Lynn Margulis has been associated: the theory of Gaia. Gaia, which only recently has emerged from the tenuous realm of scientific hypothesis, holds that the evolution of the earth and all life on it has been regulated by the action of life itself. This theory has been the subject of two books by the British atmospheric chemist James Lovelock (1979, 1988), who first developed Gaia. It is important for teachers and students to recognize that Gaia is very controversial, but the controversy merely establishes its scientific credibility. I must conclude this mention of Gaia by saying that students love it. They love being able to relate their understanding of water, oxygen, and carbon cycling, of extinction, of environmental imbalance to the existence of life on the planet.

Recently, I happened across a quote from Alan Mix, a climatologist at Oregon State University. Commenting about the uncertainties in his field, he said that "we've got lots of ideas and we're out there chasing them. We really don't know which way it's leading but that's good. It's called science" (Monastersky, 1988). This to me is the essence of scientific thought. Having ideas, investigating them, and not knowing where investigations will lead are what science is all about. This is true, whether it is in a sophisticated laboratory or in a high-school science classroom.

Let students do laboratory research in which the answer is unknown. Let them use microorganisms, including bacteria, slime molds, and algae. Many biological principles—including those related to population growth, natural selection, genetics, immunology, and physiology—can be investigated with these organisms in stimulating, open-ended activities. These kinds of experiments lend themselves to manipulation of variables, collection and organization of data, and data analysis with the computers now found in many science classrooms. Also, these organisms are easy, safe, and relatively inexpensive to use. Let the entire class design projects using vinegar eels. The results could surprise everyone!

Green plants and algae are superb organisms for classroom use. They can be exposed to a multitude of variables, including toxic substances and other environmental factors of great concern to human life today.

All this is not to say that dissection should not be a part of high-school biology. Except for the dissection of simple creatures, such as earthworms, my own preference is for dissections to be used mainly in advanced-placement biology courses by students who have already had 1 year of biology. Use living animals to investigate processes of life. Borrowing freely from Alexander Pope, "the proper study of biology—the science of life—is life."

WHAT IS THE SOCIAL IMPORTANCE OF BIOLOGY EDUCATION?

Jacob Bronowski once wrote that "men have asked for freedom, justice, and respect precisely as the scientific spirit has spread among them" (Bronowski, 1956). The spirit of science will *not* spread, unless the public perceives it as part of its world, as having genuine meaning for its life. Teachers can lecture as long as they want about how our bodies are made of billions and billions of cells, how our genes are made of DNA, and so forth and so on. We, as biologists, are fascinated by gene theory, genetics, ecology, and other biological phenomena, or we would not be teaching about them. It is *critical*, however, that we recognize that a discipline-centered curriculum may serve the needs of preprofessional science students, but not the needs of the average citizen. College curricula—taken by education students studying to be biology teachers—are structured to meet the needs of college teachers, research biologists, or future physicians. They are *not* designed to educate the average citizen.

In recent years, there has been a lot of discussion about scientific literacy, or the lack thereof, in the general populace. Today more Americans read the astrology column than news of scientific discoveries. More people have confidence in the pronouncements of Velikovsky and van Daniken than in the work of Jonas Salk or James Watson. Naive, even reckless thinking of this sort distresses science educators, but should further inspire us to find ways of making biology—the science course taken by more high-school students than any other—an experience with a lifelong, positive impact.

Morris Shamos, a former president of National Science Teachers Association, wrote that the goal of scientific literacy for all citizens would be difficult to achieve, and efforts to attain it would be counterproductive, turning off many students as they are required to learn vast arrays of facts, scientific history, and other data that have little meaning for them in any part of their lives. Instead, he said, teachers should try to foster within them an appreciation of the scientific process. Educators should allow them time for open-ended experimentation, then develop within them the necessary skills to relate science to their lives (Shamos, 1988).

A story in the October 1988 issue of the *American Scientist* brought home the need for a practical scientific literacy in this country. In San Diego last year, a stretch of Interstate 5, the major north-south route through California, was shut down for 8 hours when the report came through that a 50-pound bag of iron oxide had spilled from a truck. Finally, more than 2 hours after a crew from a hazardous-waste management company had worked for several hours to clean up the spill, someone recognized that what had spilled on the highway was no more than rust. No one had the sense to order workers to "get that rust off the road!" Is this a case of stupidity? Ineptitude? Scientific illiteracy?

It is important for all students to spend part of their class time several times a week discussing current science topics. Aside from AIDS, the topic of major scientific discussion in America in recent months has been the greenhouse effect. There is no question that there have been an extraordinary number of weather-related events the last few months. In September, Hurricane Gilbert, the "storm of the century," pounded Mexico and the coast of Texas. Bangladesh has experienced the worst flooding in its history, and fires have destroyed nearly half the forest in Yellowstone National Park. And I need not mention this summer's devastating heat and drought. Chicago broke all records for days with temperatures above 90°F—47.

Of all these events, we can be reasonably certain that only Hurricane Gilbert was not in some way influenced by human activity. Bangladesh is flooded because the mountains to the north have been stripped bare of vegetation by people seeking firewood. The slopes are no longer able to absorb and retain rainwater as they did in the past, and the people in the floodplains downstream pay the price. Yellowstone's fires are due in large part to decades of mismanagement by short-sighted people who failed to recognize that fire is an essential part of the ecology of forests. The situation in Yellowstone is fascinating and has caught the fancy and genuine interest of the entire nation. What a wonderful way for students to learn about ecology!

The greenhouse effect is particularly appropriate for classroom discussion. Science and controversy are common bedfellows, and it is easy for scientists to find evidence both *for* and *against* the existence of a greenhouse phenomenon. The jury is still out on whether increasing carbon dioxide levels caused the hot dry summer, whether ozone was a factor, and whether temperature increases will continue.

A broader spectrum of topics appropriate for use in a biology classroom includes land use (have students survey their own neighborhoods for the presence of green space), water resources (what happens to Lake Michigan affects the entire Midwest), and extinction and endangered species (students are interested in efforts to preserve the California condor, the black-footed ferret, and the great whales). Students are responsive to issues of human health and disease, energy resource management, and ethical issues involving reproduction, caring for the terminally ill, and aging. These topics are particularly useful for teaching in inner-city schools, where so many students are touched by these aspects of life and death.

These ideas all have the advantage of relevance *today*. They are biological and directly related to human existence.

As I tell my students at the beginning of the year, science *is* fun. It is discovery, it is investigating, it is asking questions. The more questions asked, the better. Yes, it is work, but it is probably the most adventurous, exciting work they will do in their high-school careers.

REFERENCES

Allman, W. E. 1988. Drowning in a sea of knowledge. U.S. News World Rep. 105(10):59.

American Scientist. 1988. Science observer: A special report on scientific literacy. Amer. Sci. 76:439-449.

Berman, W. 1984. Dissection dissected. Sci. Teach. 51(6):42-49.

Biological Sciences Curriculum Study. 1980. Biological Science: A Molecular Approach. 4th ed. Lexington, Mass.: D. C. Heath.

Biological Sciences Curriculum Study. 1987. Biological Science: An Ecological Approach. 6th ed. Dubuque, Iowa: Kendall/Hunt.

Bronowski, J. 1956. Science and Human Values. New York: Harper and Row.

Hurd, P. D., et al. 1980. Biology education in secondary schools of the United States. Amer. Biol. Teach. 42:394.

Lovelock, J. E. 1979. Gaia: A New Look at Life on Earth. Oxford, England: Oxford University Press.

Lovelock, J. E. 1988. The Ages of Gaia. New York: W. W. Norton.

Mayer, W. V., Ed. 1970. Biology Teachers' Handbook. New York: John Wiley.

Monastersky, R. 1988. Ice age insights. Sci. News 134(12):184-186.

Oxford Dictionary of Quotations. 1980. P. 269. 3rd ed. Oxford, England: Oxford University Press.

Shamos, M. 1988. The lesson every child need not learn. The Sciences July/August 14-20.

Weisskopf, V. 1984. Keynote Address. Annual Symposium for Science and Mathematics Teachers, May 14, 1984, University of Chicago.

PART IV

Instructional Procedures and Materials

17

To Weed or to Cultivate—Which?

MARY BUDD ROWE

To weed or to cultivate—which? That is the question we must ask of all that we do in biology education at whatever level it takes place. In fact, it appears that "weed," rather than "cultivate," is the dominant strategy at virtually all levels of biology instruction, at least to the end of the sophomore year in college. Biology is host to the most exciting ideas and could have more impact on the quality of student life than any other curricular offering—but you could not guess that from current textbooks or from curricular outlines or from standardized tests or from much of the observable instruction. The weeding approach appears in some places as early as the middle-school life-science course, becomes more vigorous in the high-school biology program, and goes on with a vengeance in the beginning university courses.

What would we do differently if we shifted from weeding to cultivating? For one thing, we would put story lines, or themes, or some articulated patterns of ideas back into the texts, i.e., provide some meaningful frameworks for the budding and attachment of new shoots of information. Most of the current biology texts read like glossaries. Denuded of story lines, these products of massive agglutinations of facts were induced in response to an epidemic of testing, which currently plagues the educational countryside.

Mary Budd Rowe is professor of education at the University of Florida, a former president of the National Science Teachers Association, and author of numerous papers on high-school biology education.

That means, of course, that in order to change texts we need to change tests to conform to the images, ideas, attitudes, and patterns of relationships among these that we want to characterize a program focused on cultivating a rich biological perspective in all students given into our care for roughly 150 hours in a typical academic year. One hundred fifty hours is all we have to start a mental mutation in our students.

Aside from changing the tools—i.e., texts and tests—we have to consider faculty susceptibility to the ideas of a biology with a much broader perspective than they feel free to take in the currently prevalent "weed 'em out" ideology. Under the philosophy of cultivation, we do more to help more students to achieve and maintain an interest in biology for the rest of their lives. We know something of what it would take to make that happen, but we are like the county agent who has a "new" method that will increase productivity, but can't find anyone willing to risk a change. What must we do?

Consider first some of the major questions in the minds of adolescents. What can a biology program contribute to their search for solutions? Certainly, we cannot ignore the questions. They tell you what the agenda is from the students' perspective.

- What kind of country is this?
- What values control activities?
- Where do I fit in?
- Do they expect me to succeed or fail?
- How much effort do I need to make?
- Is success worth the effort?
- Can I get help?
- Do I have the energy and endurance?
- What happens if I do not make the effort?
- What am I up against? What is the competition?
- What difference can I make?
- Do I care? Does anybody care?

Instructors and program developers also have an agenda. The agendas must be effectively meshed.

We must attend to both the content and the process by which students become engaged with the ideas of biology. The cycle of relationships can serve as a useful template for planning purposes (Figure 1). It depicts fundamental elements that ought to be addressed in a biology program.

With the template as a guide, examine the texts, tests, curriculum—instruction as it actually takes place. Ask how often in the 150 hours an opportunity to go completely around the cycle (starting at any place) occurs. In the weeding paradigm, which is largely turf-bounded, it rarely happens. Participation in such a cycle, however, is essential to the growth

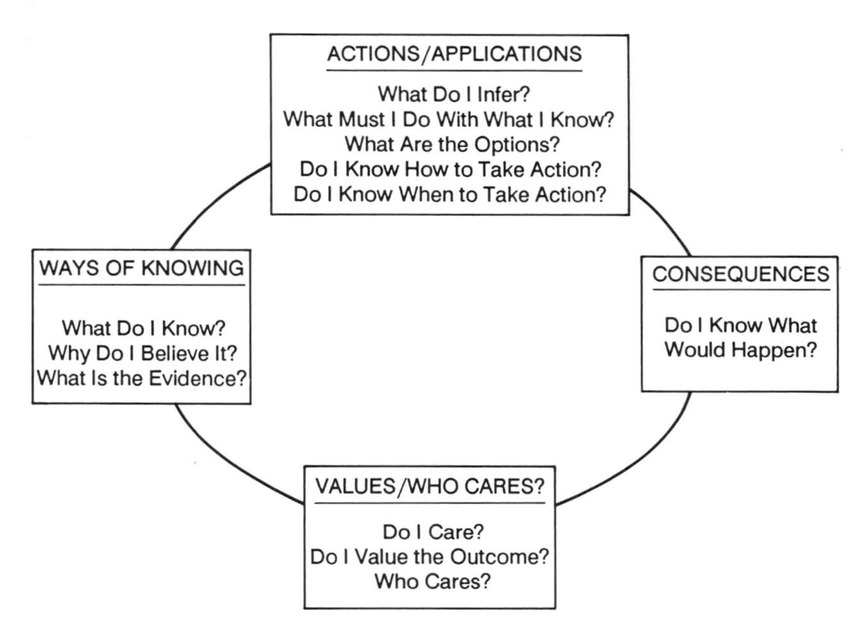

FIGURE 1 Guide for examining curriculum: fundamental elements in a program (Rowe, 1983).

of the biological perspective that we want to cultivate in students, who will be spending the larger part of their lives in the next century. The cycle carries in it the theme of connectedness, instead of the aura of chaos.

Of all the scientific and technological ideas confronting us today, possibly the most important is the recognition that humankind is a single world-wide, interdependent species. Survival, therefore, may depend on our ability to speed up the process of cooperation. That, in turn, depends on whether we can develop ways of thinking and feeling that support the process. Attitudes, beliefs, emotions, tastes, and ideologies can either motivate us to engage in productive problem-solving or turn us into fearful, turf-ridden, withdrawn people. They can energize us or enervate us. They can give license to our curiosity and fuel our persistence in the face of difficulties, or they can turn us into frenzied fanatics. They can be the source of public venturesomeness or public apathy. Our world of divergent communities is kept separate by the firmness of differing beliefs, aspirations, trusts and distrusts, convictions, and habits of resolving conflicts. These are the gatekeepers of our future, for they are the framework within which we interpret our experiences, make decisions, and take actions. Presumably, education can make a difference as to the direction in which they develop. If we regard attitudes, beliefs, feelings, tastes, and curiosity as untapped sources of national power to be cultivated in part by what we do in biology programs, then we may see our purpose well expressed by Gwen Frostic, Michigan naturalist, poet, and artist:

> We must create a great change
> in human direction—
> an understanding
> of the interdependency
> by which the universe evolves.
> Know
> —that knowing—
> is the underlying foundation
> for the life we must develop . . .
> We cannot leave it to the scientists—
> nor any form of government—
> each individual
> must fuse a philosophy
> with a plan of action.

REFERENCES

Frostic, G. 1970. Beyond Time. Benzonia, Mich.: Presscraft Papers.

Rowe, M. B. 1983. Science education: A framework for decision makers. Daedalus 112(2):123-142.

18

Biology Learning Based on Illustrations

ROBERT V. BLYSTONE

Illustrations have become an essential part of the biology learning experience. Encompassing graphs, charts, flowcharts, diagrams, line drawings, photographs, and symbols, illustrations are found in biology textbooks, computer programs, instructional audiovisual media, and even classroom wall coverings. In a world where 85% of all the messages we receive are visual (Doblin, 1980), illustrations are too often poorly used in both teaching and learning strategies. Proper development and use of illustrations can appreciably aid in the understanding and advancement of biology learning.

IMPORTANCE OF ILLUSTRATIONS TO TEXTBOOKS

Virtually every high-school biology course uses textbooks. Goldstein (1978) has estimated that 75% of the classroom time and 90% of homework time involve textbook use. The examination of biology textbooks provides a good starting point for the evaluation of the impact of illustrations on biology learning.

From the tiger on the front of *HBJ Biology* (Goodman et al., 1986) to the lurking black panther of Johnson's *Biology* (1987), the covers of

Robert V. Blystone, professor of biology at Trinity University in San Antonio, received a B.S. in 1965 from the University of Texas, El Paso, and an M.A. and Ph.D. in 1968 and 1971 from the University of Texas, Austin. He has taught at Trinity University since 1971 and served as chairman of biology in 1984-1986. Dr. Blystone's research interests include science textbooks and electron microscopy of developing lungs.

today's high-school and college biology textbooks symbolize the importance of illustrations. Frequently, biology teachers identify textbooks by the illustration on the cover: "the red-blood-cell book," *Heath Biology* (McLaren and Rotundo, 1989); "the owl book," Holt's *Modern Biology* (Towle, 1989); or "the parrot book," Mader's *Biology* (1987). Although the pictures are purely decorative, publishers willingly spend upwards of $10,000 for the perfect textbook cover picture. Many consider the colorful artwork in today's textbooks as frivolous and there primarily to sell the books (Davies, 1986). From half to three-quarters of the cost of the development of a new high-school textbook is invested in artwork and graphic design (John McClements, Addison-Wesley Publishers, personal communication). But is this emphasis on colorful covers and on the artwork on the pages in both high-school and college textbooks frivolous?

Comparison of textbook editions reveals evolutionary changes in content and format. The first edition of Keeton's *Biological Science* appeared in 1967, and the fourth edition in 1986 (Keeton and Gould, 1986). With more and larger pages, the fourth edition has over 75% more page space; however, the number of words has increased a scant 10%. The majority of the additional space has been used for additional illustrations.

Similar increases in textbook size, primarily for illustrations dealing with new concepts, may be seen in other college and high-school book editions. Three reasons contribute to this change:

- Good artwork does sell textbooks.
- Research has shown that illustrations are effective cognitive devices.
- Illustrations keep the length of a textbook down by presenting concepts in less space than text alone.

The first reason should come as no surprise. The second reason has recently come to light. Until about 20 years ago, the conclusion of most research concerning illustrations as learning devices was that they were neutral or negative in effect. With the work of Dwyer (1972), Twyman (1985), Holliday (1975), and others, a large body of evidence has been collected that properly designed illustrations do work (Levie, 1987). The third reason is not generally recognized. Blystone and Barnard (1988) showed that the average major texts will reach 1,450 pages by the year 2000 at the present rate of increase.

Publishers fully recognize that few people want a 1,450-page college introductory textbook; yet academe wants nothing left out of the book. An illustration takes less space to present complex information than does verbal text; remember that a picture is worth 1,000 words. By using more illustrations, a publisher is increasing the attractiveness of the book, increasing understandability, and saving precious space. These three reasons have contributed to illustrations' becoming more important to the message and selling of the textbook.

Textbooks for college biology majors in the 1950s averaged over 600 pages and in the 1980s over 1,100 pages long. Blystone and Barnard (1988) reported a decline in the proportion of pages in college biology textbooks with no illustrations from 52% to 22% during this period. The use of photographs increased nearly threefold during the same interval. In contrast with today's textbooks, no color was used in the college biology books of the 1950s. The changes reported for college biology textbooks are mirrored in the high-school texts. The 1987 edition of *BSCS* (the green edition) is over 1,000 pages long. The emphasis on more illustration in high-school textbooks has also carried over from the college field. The 1989 edition of *Modern Biology* (Towle, 1989) has nearly all color artwork, and even many of the scanning and electron micrographs are colorized.

INCREASED COMPLEXITY OF ILLUSTRATIONS

The subject and content of college textbook illustrations have changed considerably over the last 30 years. Figure 1 describes the changes in illustration subject emphasis in college textbooks. On the basis of five levels of biological organization for subject identification, illustrations with molecular and cellular content have appreciably increased. Illustrations concerning whole organisms have also increased. However, the frequency of illustrations with organ- and tissue-level subjects has remained nearly the same (Blystone and Barnard, unpublished). The increase in cellular and molecular illustrations is predictable, given the prominence of today's genetic and biochemical information. The data on the other three levels are difficult to explain.

The range of complexity of illustrations can be demonstrated by comparing the first and fourth editions of Keeton's *Biological Science*. On page 291 of the first edition, a two-dimensional, black-and-white rendering of a grasshopper is seen. The drawing occupies a quarter of a page, in contrast with an eighth-of-a-page, three-dimensional, color version of a grasshopper seen on page 366 in the fourth edition. The latter illustration is far more realistic than the original, and it uses space more economically. This is a common strategy in textbook revisions.

Comparisons of fluid mosaic membrane models from Merrill's *Biology: Living Systems* (Oram, 1983, p. 75), *HBJ Biology* (Goodman et al., 1986, p. 102), and *Heath Biology* (McLaren and Rotundo, 1985, p. 71) reveal great variation in complexity of the representation. In terms of content, the *Heath Biology* high-school textbook illustration would rival Johnson's *Biology* (1987, p. 57), a college rendering of the same topic.

Variation in complexity of illustrations is apparent in cell models. Four

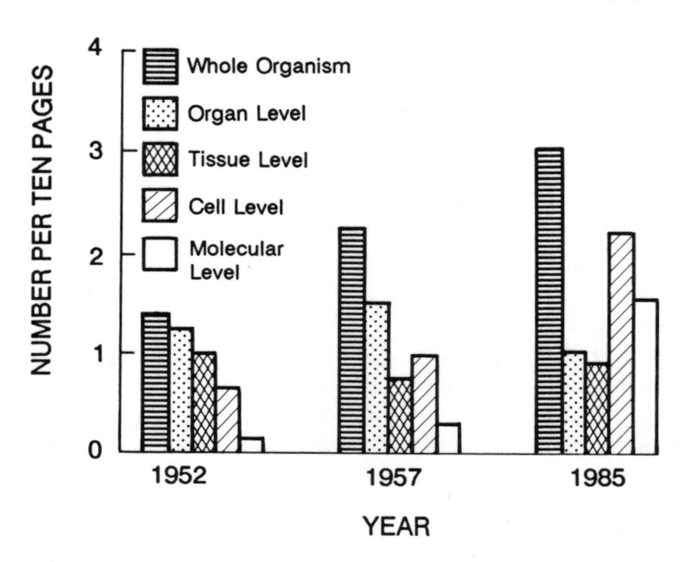

FIGURE 1 Subjects of textbook illustration.

examples would be *HBJ Biology* (Goodman et al., 1986, p. 83), Addison-Wesley's *Biology* (Kormondy and Essenfeld, 1988, p. 111), Holt's *Modern Biology* (Otto and Towle, 1985, p. 60), and BSCS *Biological Science: A Molecular Approach* (1985, p. 25). The simplest is the HBJ diagram, which reiterates the classic cell diagram in the *Scientific American* (1961) special issue on the cell. The most complex would be the Addison-Wesley illustration. All four texts are oriented toward the same market; yet, the illustrations vary dramatically in complexity. Which model would best suit the cognitive level of the audience? The answer is left wanting.

Diagrams in textbooks today at both high-school and college levels are more complex. They show much more motion, events at different times within the same illustration, spatial relationships, and process. But an evenness of presentation in terms of this complexity does not exist between textbooks and even within the same textbook. In a period when verbal text complexity is constrained by reading level and interest formulas, the illustration levels can swing like the moods of a manic-depressive.

SUBTLETIES OF ILLUSTRATION DESIGN

Analysis of illustrations can reveal many subtleties. For example, compare the endoplasmic-reticulum (ER) illustration in the 1985 *Heath Biology* (McLaren and Rotundo, 1985, p. 73) and Addison-Wesley's *Biology* (Kormondy and Essenfeld, 1988, p. 106). The Heath version shows ribosomes

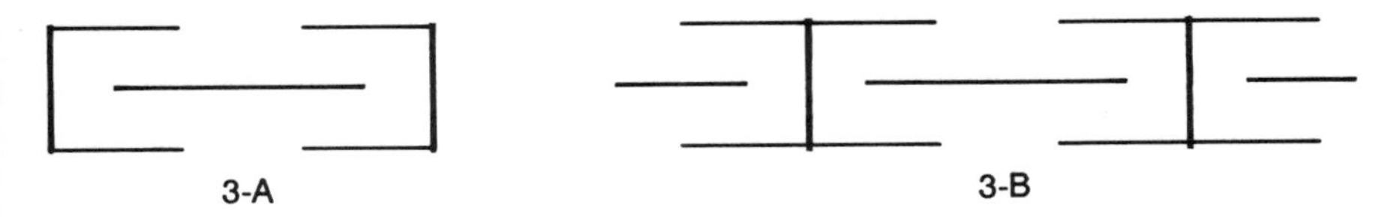

FIGURE 2 Different presentations of sarcomere concept.

in random array on the ER surface, whereas Addison-Wesley depicts ribosomes beaded on messenger RNA and attached to the ER.

A more subtle problem can be found when comparing the sarcomere illustrations of Keeton and Gould's *Biological Science* (1986, p. 541) and Raven and Johnson's *Biology* (1986, p. 132) with that of Curtis's *Biology* (1983, p. 810) and Mader's *Biology* (1987, p. 548). Figure 2 is a recreation of these diagrams. The first two texts are represented on the left side of Figure 2, and the latter two on the right side. Obviously, the left side shows one sarcomere, and the right side one sarcomere and portions of two adjacent sarcomeres.

On a recent biology advanced-placement examination, where muscle contraction was explored in a question, most students using illustrations like the left side of Figure 2 would present sarcomere structure as did their text—exactly one sarcomere. In the essay component of the answer, many of these students did not indicate that many sarcomeres operate together. These students viewed muscle contraction in terms of exactly one sarcomere—just as the illustration presented sarcomere structure. Students who learned with sarcomere textbook illustrations like the right side of Figure 2 more frequently recognized that many sarcomeres operate together during contraction. All diagrams are correct in this case, but the students received a better perspective of muscle contractions with the two-sarcomere illustration.

These subtle distinctions in illustrations can affect how a student learns. Recall again the model cell diagram from the September 1961 *Scientific American*. This cell model, now known to be incorrect, still influences learning today. Illustrations often outlive the corresponding text. Illustrations frequently make a more lasting impression on learning than does the verbal text. Subtle design features, such as beaded ER ribosomes and two-sarcomere models, have lasting effects to an extent greater than many people are generally aware.

CURRENT DIRECTION OF ILLUSTRATIONS

Textbook graphic designers are taking advantage of new computer technology. High-school biology books are pioneering the use of colored images; colored gene sites are on colored bacterial DNA in Holt's *Modern Biology* (Towle, 1989, p. 7). College textbooks are beginning to emulate the high-school books in this practice of adding color to both electron and light micrographs (Campbell, 1987, pp. 540-541).

Dramatic use of color and graphic layout can be seen in the illustrations of the spider and the evolution of the brain in Holt's *Modern Biology* (Towle, 1989, pp. 487, 625). The autonomic-nervous-system wiring diagram in *Heath Biology* (McLaren and Rotundo, 1989, p. 701) is far more dramatic than a similar illustration in *Gray's Anatomy* (Warwick and Williams, 1973, p. 1066). The Heath diagram also matches the medical-school text in content. Many images in today's biology textbooks are constructed in such a way as to make them comparable with commercial messages.

Holt's *Modern Biology* (Towle, 1989, p. 52) includes a computer-generated illustration of protein structure showing molecular domains. This sophisticated illustration appears in a feature on proteins. The illustration is only decorative; however, the student reader is never told how to read it. Unlike the viewer of the decorative illustrations of the past, the reader does not have the experience needed to interpret the computer image. Tigers on a book cover may be one thing; computer-produced protein molecular domains are quite another. Publishers are trying very hard to make their textbooks look current. The protein-molecule illustration in Holt is current—so much so that the reader has to be told what it means. This tendency of using illustrations, decorative and otherwise, that have no footing in the realm of common experience is increasing.

Depicting concepts through the use of stylized illustrations has become common. People who produce and use books take it for granted that a student will understand the stylized cellular metabolism sequence in an illustration like that in *Heath Biology* (McLaren and Rotundo, 1989, p. 96) or that a student will understand the four-step packing sequence of DNA into a chromosome as presented in *Heath Biology* (p. 171). A definite trend in textbook illustration is assuming that students have sufficient experience in reading illustrations, just as the same students presumably know what the X and Y coordinates are on a graph and on which axis the independent variable should be placed. Those who teach know one cannot make these assumptions about high-school students' graph-reading skills. Publishers should know that some decorative illustrations now coming into use are beyond the comprehension of the intended audience. These illustrations are appearing in textbooks for the sake of being current.

This problem exists at the college level, too. Keeton and Gould's *Biological Science* (1986, p. 580) breaks new ground by showing a computer-generated, three-dimensional graph of magnetic fields and their relation to bird navigation. The diagram calls for interpretative sophistication on the part of the college student.

TABLE 1 Illustration Types and Characteristics

Picture Type	Function	Purpose	Effectiveness
Decorative	Contributes appeal	Increased attention span	Little or none
Representational	Adds concreteness	Concrete visual representation; facilitation of memory	Moderate
Organizational	Adds coherence	Thematic organization; facilitation of memory	Moderate to substantial
Interpretational	Adds comprehensibility	Understanding; facilitation of memory	Moderate to substantial
Transformational	Provides a retrieval system	Direct impact on memory	Substantial

The current direction of illustration is toward increasing complexity; this increase is being speeded by computer graphic technology. Interpretation of these images calls forth the need for increased visual literacy by both the student *and* the teacher.

IMPORTANCE OF VISUAL LITERACY

If illustrations are so influential in learning, what degree of complexity should they have? One would argue that generally the more accurate a diagram, the more complex it will be. A good illustration is very much like good prose: difficult to produce. Far more people are able to recognize good prose than a good illustration. Considerable effort is expended in teaching verbal literacy, but little toward visual literacy.

A variety of schemata have been described for categorizing illustration types. I will explore only one of the simpler examples. Levin et al. (1987) grouped illustrations into five types: decorational, representational, organizational, interpretational, and transformational. Table 1 indicates the function, purpose, and effectiveness of each category.

This list is presented as a challenge to the reader to consider into which category an illustration falls. This is a step toward visual literacy. By being able to categorize illustrations into functional groups, a teacher can determine how best to use an illustration in a class, or a reviewer of a textbook can determine how well the illustrations have been incorporated into the book. Does the text ask the illustration to perform the proper function?

MAXIMIZING ILLUSTRATION EFFECTIVENESS

Several steps can be taken to maximize the effectiveness of illustrations. Cellular and molecular illustrations can be among the most complex of contemporary diagrams. These models represent magnifications of 1,000,000 times or more. An example of this type of illustration would be that in Becker's *The World of Cells* (1986, p. 329). This figure illustrates seven distinct periods and at least 13 discrete events in producing a protein. One approach to helping students cope with the informational content of such a diagram is to allow them to verbalize the illustration. By translating the visual image back into text, the student will learn more.

Another approach is to allow the student to build a model that represents the illustration. Doblin (1980) indicated that model-building was the most realistic form of illustration. Our experience confirms this view. When we give students clay, pipe cleaners, toothpicks, and foam dinner plates, modeling the seven-step process of protein manufacture involving ER is possible. Each step taken by a student in building the model reveals that student's understanding of the process. The student can be challenged as to why the model was manipulated in the way that it was. Model-building with complex textbook illustrations is an excellent way to pace a student's learning. Often these models can be built with a minimum of expense.

In an age when children can rotate toys into numerous positions to create different objects, it is time to provide scientific toys that can transform amino acids to proteins. With imagination, scientific illustrations can serve as a template for a whole new generation of toys. Imagine an electron-cascade game that re-enacts the energy flow represented in a mitochondrial membrane illustration. Must educational toys be only in the realm of preschoolers?

With simple verbalization and model-building, illustrations can be augmented into even more powerful learning tools. Perhaps publishers can consider these augmentations when designing illustrations and resource materials for texts.

SUGGESTIONS FOR THE FUTURE

It is important that biology textbook users realize that illustrations have far more important roles than decoration. Too few people recognize that illustrations are being used by publishers and authors to keep down textbook length. Also, illustrations are being used commercially to indicate how current a book is. Computer-aided advanced graphic design gives both high-school and college textbooks a technological feel. This technological feel implies a sense of being up to date. Also, these new graphics allow ambitious presentations not possible before. Some illustrations challenge the visual literacy of the reader.

Complexity of textbook illustrations has been a concern of this paper. But illustration can use other media. Almost 100% of the nation's public high schools have computers (Gibbons, 1988). Often, these computers are accompanied by software with computer-assisted drills in biology subjects. Usually, these drills have illustrations associated with the presentation. These illustrations vary a great deal in their complexity. Better computer systems can rotate molecules and add animation to illustrations. With CD-ROM and videodisk technology, illustrations can have even wider implications in learning. (See Buddine and Young, 1987, for an explanation of this new technology.)

Computer-based illustrations offer opportunities not possible with textbook-based illustrations. The traditional textbook illustration has only recently been verified as being instructionally significant. Now evaluation systems must be worked out to determine how these new media influence illustration-based learning. Even more exciting is the possibility of putting holograms into textbooks.

In conclusion, effective use must be made of textbook illustrations. These illustrations must not be taken for granted. With the increasing complexity of illustrations, students must be aided in interpreting the new generation of visual information. Publishers and teachers alike have the obligation of bringing to students images that can be understood.

REFERENCES

Becker, W. 1986. The World of Cells. Menlo Park, Calif.: Benjamin/Cummings.

Blystone, R. V., and K. Barnard. 1988. The future direction of college biology textbooks. BioScience 28(1):48-52.

BSCS (Biological Sciences Curriculum Study). 1985. Biological Science: A Molecular Approach. Lexington, Mass.: D. C. Heath. (the Blue Edition)

BSCS (Biological Sciences Curriculum Study). 1987. Biological Science: An Ecological Approach. 6th ed. Dubuque, Iowa: Kendall/Hunt. (the Green Edition)

Buddine, L., and E. Young. 1987. The Brady Guide to CD-ROM. New York: Prentice Hall.

Campbell, N. A. 1987. Biology. Menlo Park, Calif.: Benjamin/Cummings.

Curtis, H. 1983. Biology. 4th ed. New York: Worth Publishers.

Davies, M. J. 1986. Making kids read junk. Curriculum Review 26(2):11.

Doblin, J. 1980. A structure of nontextual communications, pp. 89-111. In P. A. Kolers, M. E. Wrolstad, and H. Bouma, Eds. Processing of Visible Language 2. New York: Plenum Press.

Dwyer, F. M. 1972. The effects of overt responses in improving visually programed science instruction. J. Res. Sci. Teach. 9(1):47-55.

Gibbons, J. H. 1988. Power ON!: New Tools for Teaching and Learning—Summary. Washington, D.C.: U.S. Government Printing Office. (GPO. No. 052-003-1125-5).

Goldstein, P. 1978. Changing the American School Book. Lexington, Mass.: Heath.

Goodman, H. D., T. C. Emmel, L. E. Graham, F. M. Slowiczek, and Y. Shechter. 1986. HBJ Biology. Orlando, Fla.: Harcourt Brace Jovanovich.

Holliday, W. G. 1975. The effects of verbal and adjunct pictorial-verbal information in science instruction. J. Res. Sci. Teach. 12(1):77-83.

Johnson, L. G. 1987. Biology. 2nd ed. Dubuque, Iowa: W. C. Brown Publishers.

Keeton, W. T. 1967. Biological Science. New York: Norton.

Keeton, W. T., and J. L. Gould. 1986. Biological Science. 4th ed. New York: Norton Publishers.

Kormondy, E. J., and B. E. Essenfeld. 1988. Biology: A Systems Approach. Menlo Park, Calif.: Addison-Wesley.

Levie, W. H. 1987. Research on pictures: A guide to the literature, pp. 1-50. In D. A. Houghton and E. M. Willows, Eds. The Psychology of Illustration. Vol. 1—Basic Research. New York: Springer-Verlag.

Levin, J. R., G. J. Anglin, and R. N. Carney. 1987. On empirically validating functions of pictures in prose, pp. 51-85. In D. A. Houghton and E. M. Willows, Eds. The Psychology of Illustration. Vol. 1—Basic Research. New York: Springer-Verlag.

Mader, S. S. 1987. Biology: Evolution, Diversity, and the Environment. 2nd ed. Dubuque, Iowa: W. C. Brown.

McLaren, J. E., and L. Rotundo. 1985. Heath Biology. Lexington, Mass.: D. C. Heath.

McLaren, J. E., and L. Rotundo. 1989. Heath Biology. Lexington, Mass.: D. C. Heath.

Oram, R. F. 1983. Biology: Living Systems. 4th ed. Columbus, Ohio: Merrill.

Otto, J., and A. Towle. 1985. Modern Biology. New York: Holt Rinehart & Winston.

Raven, P. H., and G. B. Johnson. 1986. Biology. 2nd ed. St. Louis, Mo.: Times Mirror/Mosby College.

Scientific American. 1961. The Cell Issue. 205(3):1-304. (The Living Cell: Readings from Scientific American. 1965. San Francisco, Calif.: Freeman.)

Towle, A. 1989. Modern Biology. Austin, Tex.: Holt Rinehart & Winston.

Twyman, M. 1985. Using pictorial language: A discussion of the dimensions of the problem, pp. 245-312. In T. M. Duffy and R. Waller, Eds. Designing Usable Texts. New York: Academic Press.

Warwick, R., and P. L. Williams. 1973. Gray's Anatomy. 35th British ed. Philadelphia, Pa.: Saunders.

19

Teaching High-School Biology: Materials and Strategies

RODGER W. BYBEE

WHOM ARE WE TEACHING BIOLOGY?

High-school biology is offered in 99% of high schools in the United States (Weiss, 1987). This is a 4% increase since 1977 (Weiss, 1978). Biology is the most commonly offered science course—35% of all science courses. Half of all science classes relate to the biological sciences (Weiss, 1987). It is safe to say that biology is taken by the majority of high-school students. And for many of those students, biology is the last science course they will take.

It is absolutely essential to consider the demographics of education as we look for a reform of biology education. In *All One System*, Harold Hodgkinson (1985) presents demographic trends—changes in population groupings that move through the educational system. Hodgkinson summarized his findings (p. 7):

> What is coming toward the educational system is a group of children who will be poorer, more ethnically and linguistically diverse, and who will have more handicaps that will affect their learning. Most important, by around the year 2000, America will be a nation in which one of every three of us will be non-white. And minorities will cover a broader socio-economic range than ever before, making simplistic treatment of their needs even less useful.

Rodger W. Bybee is associate director of the Biological Sciences Curriculum Study (BSCS) in Colorado Springs. Before joining BSCS, Dr. Bybee was associate professor of education at Carleton College. He is principal investigator for the new BSCS elementary-school program, Science for Life and Living: Integrating Science, Technology, and Health.

Other national reports serve to remind us that our educational programs at the precollege level must recognize the personal needs of *all* youth and the aspirations of society. One such report is *The Forgotten Half: Non-College Youth in America* (Commission on Work, Family, and Citizenship, 1988). This report is a counterpoint to the numerous reports that explicitly or implicitly focus on the college-bound student.

Whom are we teaching biology? We are teaching the majority of students. And we must recognize that the majority is a diverse group, with different needs, perceptions, and aspirations. High-school biology should be designed for all students, those who are college-bound and those who will enter the workforce immediately after high school.

CHARACTERISTICS OF STUDENTS

Contemporary research findings about students as learners underlie my discussion of instruction. One finding is that students are motivated to learn science. They are naturally curious about all aspects of the biological world. Whether it is recognizing plants and animals, understanding biotechnology, or investigating ecological systems, students have an interest in their world and seek explanations for how things work.

A second finding is that students already have explanations, attitudes, and skills when a biology lesson begins. Students' explanations, attitudes, and skills may well be inadequate, incomplete, or inappropriate. Contemporary educational researchers use such terms as "misconceptions" and "naive theories" to characterize the cognitive component of student understanding. Briefly, students interpret instructional activities in terms of what they already know; then they actively seek to relate new concepts, attitudes, or skills to their prior set of concepts, attitudes, or skills. The assimilation of new experiences is based on the students' prior experiences, and it may or may not get "learned" the way the teacher intended. Students' learning is accurately viewed as the process of refining and reconstructing extant knowledge, attitudes, and skills, rather than the steady accumulation of new knowledge, attitudes, and skills.

A third finding is that students have different styles of learning. "Learning style" refers to the way individuals perceive, interact with, and respond to the learning environment. Learning styles have cognitive, affective, and physical components. While instructional strategies vary between and within projects, they are based on the idea that learning style is an aspect of students' learning and should be recognized in the strategies of teaching.

The fourth finding is that students pass through developmental stages and that tasks influence learning. In the 1960s and 1970s, Jean Piaget's theory was popular, and it influenced curriculum development. Piaget's work concentrated on cognitive development. Current research in the cognitive sciences is, in many respects, an extension of Piaget's theories. Contemporary curriculum development holds a larger view of student development. In addition to cognitive development, we should also attend to the student's ethical, social, and psychomotor development. This broader view of development is important to the selection of instructional methods.

The general view of student learning presented in the four findings is constructivist. In the constructivist model, students reorganize and reconstruct core concepts, or intellectual structures, through continuous interactions with their environment and other people. Applying the constructivist approach to teaching requires the teacher to recognize that students have conceptions of the natural world. Those may be inadequate and need further development. Curriculum developers can design materials and teachers can use strategies so that students encounter objects or events that focus on the concepts, attitudes, or skills that are the intended learning outcomes. Then they can have students encounter problematic situations that are slightly beyond their current level of understanding or skill. The instructional approach then structures physical and psychological experiences that assist in the construction of more adequate explanations, attitudes, and skills. These new constructions are then applied to different situations and tested against other constructions used to explain and manipulate objects and events in the students' world. Briefly, the students' construction of knowledge can be assisted by using sequences of lessons designed to challenge current conceptions and by providing time and opportunities for reconstruction to occur.

WHAT SHOULD WE TEACH?

Through most of time, the immense journey of biological evolution has been directed by natural laws. With scientific and technological advances, such as the discovery of DNA and the development of biotechnology, and with the problems of population, resources, and environments—such as famine, destruction of tropical rain forests, and ozone depletion in the upper atmosphere—we have abilities and influences that go beyond our meager understanding and myopic visions. Evolution may now be directed by humans themselves. Here is a clear and profound connection between biology as a pure science and the influence of biology on our global society. Students need an ecological perspective. All other arguments for a particular curriculum emphasis in biology pale in comparison.

A recent editorial in *Science* (Koshland, 1988) described the importance of ecological understanding:

> Ecology, the study of the delicate balance between species and environment . . . shows that evolution has developed clever strategies . . . to use

> resources to maximum effectiveness. Those strategies sometimes involve symbiosis, sometimes tacit agreements on territory, and sometimes murderous aggression, but all are based on the assumption that resources are limited so that the clever and the parsimonious will gain relative to the inefficient and wasteful.

At the end of the editorial, Koshland made a clear connection to human populations:

> Most species struggle to overcome poverty of resources and occupy niches that allow a critical number to survive in competition with other species. Modern civilization has upset that process so that many (although certainly not all) humans are living far beyond a survival level. The brain that allowed that situation needs now to curb a primordial instinct to increased replication of our own species at the expense of others because the global ecology is threatened. So, ask not whether the bell tolls for the owl or the whale or the rhinoceros; it tolls for us.

This powerful statement has the implied theme of educating the public about global ecology. The public has an increased awareness and concern related to interactions among individuals, groups of individuals, and the environment. Public attention is directed to these primary units of ecological study. This attention has influenced the growing public concern for ecology and public debate about policies that extend the concern to human ecology.

In biology education, there has been an essential tension between the need to teach "real biology"—the science of life—and the need to achieve educational goals related to personal development and societal aspirations—the science of living. The continuing debate about the primary goals—whether the biology curriculum ought to be a science of life or a science of living—is essential to the continued evolution of biology education. The history of this debate has been described elsewhere (Rosenthal and Bybee, 1987, 1988). I perceive the contemporary resolution of the debate to favor human ecology, which should be the conceptual framework for the curriculum in biology.

The teaching of human ecology is an integrative endeavor among humanists, social scientists, and natural scientists. Separate disciplines—such as biology, sociology, psychology, anthropology, economics, philosophy, theology, and history—evolved to improve understanding of the human condition and, we may assume, the human predicament. Now, when problems cut across these disciplines, there is reluctance to transcend the disciplinary boundaries. Such reluctance must be overcome for the very reasons for which disciplines were invented—the cause of human understanding, if not survival. The idea of cooperation among the various disciplines serves to maintain the integrity of disciplines while permitting study of the unifying conceptual schemes of biology and the central issues of human ecology—population dynamics, growth, resource use, environmental practices, and the complex interaction of human populations, resources, and environment (Moore, 1985; Ehrlich, 1985).

TEXTBOOKS

To say that generally the biology textbook is the organizing framework for the curriculum and reading the textbook is the dominant method of instruction is not an overstatement. Over 90% of science teachers use published textbooks (Weiss, 1978, 1987). And science instruction tends to be dominated by teacher lectures and reading of the textbook (Weiss, 1987; Mullis and Jenkins, 1988). Any consideration of reforming high-school biology must examine the role of the textbook in instruction.

There is a contradiction associated with the use and review of textbooks. A majority (76%) of science teachers in grades 10-12 do *not* consider textbook quality to be a significant problem (Weiss, 1987). On the other hand, many educators *do* consider textbook quality and usability to be problems (Muther, 1987; Carter, 1987; AAAS, 1985; Apple, 1985; Armbruster, 1985; Moyer and Mayer, 1985; McInerney, 1986; Rosenthal, 1984).

Science teachers are clearly satisfied with the quality of textbooks. In a national survey of science education, Weiss (1987) asked several specific questions about the quality of science textbooks. Some of the items that received favorable ratings by a majority of respondents are the following:

- Have appropriate reading level (87%).
- Are interesting to students (52%).
- Are clear and well organized (85%).
- Develop problem-solving skills (61%).
- Explain concepts clearly (74%).
- Have good suggestions for activities and assignments (74%).

Why are the teachers satisfied? The textbooks are meeting teachers' needs and their conceptions of good biology and appropriate biology education. The problem here is similar to that of the biology student who has misconceptions about the energetics of cells or the mechanisms of evolution. The means of changing the misconceptions is likewise similar. There is need to challenge current concepts and introduce biology teachers to perceptions about textbooks that are counter to their own. Then, provide time, opportunities, and examples that allow teachers to reform their ideas.

We may also have to consider the questions that probe beyond those asked in the survey. For instance, the material is clear and well organized; but should we be teaching that material? Or, the textbooks develop problem-solving skills; but which problem-solving skills, and are they really developed? The problem of teacher satisfaction with textbooks is central to any reform of biology education.

Content and pedagogy are central to the textbook situation. One assessment of content is the copyright date of textbooks in use. Seventy-one percent of science classes in grades 10-12 use books with a copyright date before 1983, and 22% before 1980. So one dimension of the content problem is that the information is dated.

Gould (1988) published "The Case of the Creeping Fox Terrier Clone," in which two themes were developed. One was the presentation of controversial issues, such as evolution, in textbooks. The second, and more important, was that textbooks in a given market, like tenth-grade biology, are very similar to one another. Gould did an informal review of biology textbooks and had this to say (1988, p. 19):

> In book after book, the evolution section is virtually cloned. Almost all authors treat the same topics, usually in the same sequence, and often with illustrations changed only enough to avoid suits for plagiarism. Obviously, authors of textbooks are copying material on a massive scale and passing along to students illconsidered and virtually xeroxed versions with a rationale lost in the mists of time.

At the end of the article, Gould remarked on the educational effect of cloning (p. 24):

> [Textbook cloning] is the easy way out, a substitute for thinking and striving to improve. Somehow, I must believe—for it is essential to my notion of scholarship—that good teaching requires fresh thought and genuine excitement and that rote copying can only indicate boredom and slipshod practice. A carelessly cloned work will not excite students, however pretty the pictures. As an antidote, we need only the most basic virtue of integrity—not only the usual, figurative meaning of honorable practice but the less familiar, literal definition of wholeness. We will not have great texts if authors cannot shape content but must serve a commercial master as one cog in an ultimately powerless consortium with other packagers.

What about pedagogy? The design of textbooks supports the science teachers' increased use of lecture and decreased use of laboratory (Weiss, 1987). One can imagine the situation getting worse, because the feedback within the system will continue to support the trend. More information is added to textbooks, but teachers have a fixed time to cover information. Fewer laboratory experiments are done, because more time is needed for lectures. Somehow, the cycle must be interrupted.

Reforming the content and pedagogy of textbooks is a complicated and complex proposition. Who is in control? Authors? Publishers? State adoption committees? Curriculum developers? Administrators? Teachers? The fact is that all groups are in some control and to some degree controlled. Most of the feedback in the system tends to perpetuate the current situation. It will take the concerted efforts of those within the system to bring about change. How might this happen? We need only look back 30 years to find a historical example. Support for several innovative biology programs, such as those developed in the late 1950s and 1960s, could bring about some change. Those programs incorporated the best scientists and teachers in the design of new textbooks. The original development and field-testing of materials was heavily supported and unencumbered by restraints of the market, adoption committees, and administrative budgets. The science-education community united to develop innovative programs; then the market adapted.

What should we do differently in the 1980s? First, I think several different groups should be developing biology programs. While the Biological Sciences Curriculum Study (BSCS) was successful in developing three programs, I think there is need for even more diversity. Second, the projects should be funded by both private and public sources. The reasons for this are to encourage greater diversity and innovation of programs and to provide enough funding for significant innovation, such as the integration of technology (educational software), and major field-testing of the programs. Third, only publishers that are willing to give control of content and pedagogy to the developers should be involved in the projects, and those publishers should be involved throughout the development process. Fourth, development should include implementation of the program. Finally, teacher education at the preservice level should be integral to development and implementation of the new programs.

TECHNOLOGY

The use of educational technology has great potential for improving instruction in biology. According to Weiss (1978), computer use increases with grade levels, with approximately 36% of science classes in grades 10-12 using computers. Although the amount of time computers are used is small, at grades 10-12 computers are used primarily for drill and practice, for simulations, for learning content, and as laboratory tools (Weiss, 1987). In contrast to 1977, the 1985-1986 national survey indicated that computers are a part of science education. I assume that the trend toward increased use of computers will continue. Among the justifications for greater use of computers are the demands of an increasingly information-oriented and technological society and use of computers in the workplace (Ellis, 1984).

There have not been sufficient quantities of good software and affordable hardware for computers to have a widespread impact on curriculum and instruction in biology. Individual pieces of software are used as supplements to instruction. But the occasional application of a tutorial or simulation is not enough to bring about the reformation of thinking required to incorporate computer technologies fully into the biology program. As hardware and software evolve, there is reason to believe that they will become integral components of biology education (R. Tinker, unpublished manuscript).

There are three types of software that have immediate and important implications for instruction in biology: HyperCard, microcomputer-based laboratories, and modeling.

Hypercard

Textbooks have reached the point of diminishing returns relative to the amount of information they can reasonably contain for high-school biology. HyperCard is an educational technology that has relevance for the problem of teaching students how to ask questions and get information on selected subjects. They can simply view the information that someone else has organized, or they can "collect" information and organize it in a notebook (Kaehler, 1988).

Biology teachers are concerned that students must "learn" information that teachers do not have time to teach. HyperCard allows the students to gain access to information when they need it, to the depth that they want.

Microcomputer-Based Laboratory (MBL)

MBLs permit the acquisition of data in the laboratory through probes and sensors linked with a computer. This educational application was pioneered by Robert Tinker at Technical Education Research Centers. Data types that might be used in biology instruction include temperature, sound, light, pressure, distance measurement, electrical measurements (such as resistance and voltage), and physiological measurements (such as heart rate, blood pressure, and electrodermal activity).

MBL offers extensions of many current laboratories in biology education. It has several educational advantages, such as immediate feedback for students, capability for long-term collection of data, and easy construction of graphs for display of data. There is little reason not to use this technology in biology instruction.

Models and Simulations

Modeling tools are available in software packages that assist students in quantitative assessment. STELLA is the archetype of this software (Tinker, unpublished manuscript). Modeling applies very nicely to such subjects as population growth, resource depletion, and environmental degradation. Simulations provide students with opportunities to try ideas, change variables, and run hypothetical experiments. Computer technology affords the opportunity for students to investigate topics that they ordinarily could not study.

TEACHING

My discussion of teaching is divided into two sections. The first concerns the laboratory and the second argues for a more systematic approach to instruction. The 1985-1986 national survey indicated that since 1977, science teachers have increased the amount of time in lecture and decreased the time in laboratory activities (Weiss, 1987). There is a need to renew and expand the emphasis on the laboratory and inquiry strategies (Costenson and Lawson, 1987).

Human Ecology and the Biology Laboratory

Human ecology is the conceptual orientation that I recommend for the biology laboratory (Bybee, 1984, 1987). Human ecology as a specific approach to the laboratory is described in Bybee et al. (1981). The following are characteristics of a laboratory program with a human ecological approach. The characteristics describe an orientation and direction for the science laboratory. Table 1 compares traditional and human ecological approaches to the science laboratory.

Study of Significant Problems

Laboratory activities will be related to problems in the human environment. Problems arise from situations that involve a question, discrepancy, or decision concerning the student, society, or the environment. Investigations should be selected that provide opportunities for students to help to define problems significant to them—problems that they think they can and are willing to help to solve (Bybee et al., 1980). Investigations should be oriented toward ways of acquiring information and using that information in making decisions about current personal and social problems. The following subjects could form the basis for study: world hunger and food resources, population growth, air quality and atmosphere, water resources, war technology, human health and disease, energy shortages, land use, hazardous substances, nuclear reactors, extinction of plants and animals, and mineral resources. The selection of subjects is based on surveys of different populations, including American citizens (Bybee, 1984) and science educators in other countries (Bybee, 1987).

Study of Ecosystems

An instructional orientation toward the ecosystem is appropriate. Of necessity, biology teachers will have to include other levels of biological organization, but students can experience and understand many changes in ecosystems, especially as they study them at local levels.

TABLE 1 Comparison of Traditional and Human Ecological Approaches to Science Laboratory

Traditional Laboratory Approach	Human Ecological Approach
Students verify knowledge presented in textbook.	Students study problems involving scientific concepts and skills.
Problems are within a single scientific discipline.	Problems require an integration of disciplines.
Problems have a single cause-effect relationship.	Problems are multicausal.
A conclusion based on the data is a major component of the activity.	An interpretation of data leading to a decision is a major component of the activity.
Students use reductive methods.	Students use reductive and holistic methods.
The laboratory is primarily a classroom-based activity.	The laboratory is classroom-, school-, and community-based.
Ethics and values are not generally included.	Ethics and values are part of the decision-making process.
Experience is related to the abstract world of science.	Experience is related to the concrete world of the student.
Problems are easily defined and predictable.	Problems have undefined dimensions and unpredictable results.
Scientific concepts are studied as the "structure of a discipline."	Scientific concepts are studied in the resolution of science-related social issues.
Laboratory work is presented as short-term accumulation of data and the scientific process.	Laboratory work is presented as both short- and long-term accumulation of data and the scientific process.

An ecosystems perspective is a good way to integrate various disciplines; it provides a common conceptual framework and language. The perspective could be introduced early in the biology program and thus provide concepts and terminology for the students' continuing study.

Holistic Methods of Study

Ecologists use holistic perspectives in scientific inquiry. Holistic methods can develop the students' ability to identify various interacting parts of systems (subsystems) and to understand the behavior of whole systems. Holistic methods of study are complementary to reductionistic methods, and students should experience the appropriate application and unique strengths of these methods.

Integrative Study

Biology education has held as important goals the development of and the ability to use biological concepts and methods of biological investigation. An orientation toward human ecology expands these goals in an effort to understand and resolve human problems. Human ecology provides experience in decision-making as a means to help students contribute to the eventual amelioration of problems. Decision-making implies some understanding of the social, political, and economic realms, as well as ethics and values. The primary emphasis of biology education programs should be on the concepts and processes of biology and biological investigation. A secondary emphasis is on the application of other disciplines in the cause of understanding and resolving problems.

Development and Learning

Instruction reflecting a human ecological approach should reflect an understanding of students as learners. Obviously, a global perspective of problems related to such issues as population growth or food resources is beyond the grasp of younger children. But local problems and some basic concepts—such as the difference between arithmetic growth and exponential growth—are not too complex for young children. Successful laboratory instruction in human ecology requires recognition of students' cognitive development and learning limitations.

Perspectives of Space, Time, and Causal Relations

Laboratory experiences should expand students' perspectives of space, time, and causal relations. Over the school years, students should extend their ideas of space from local to regional to national to global perspectives. Their ideas of time should extend to the distant past and to the future. Causal relations should extend from simple cause and effect to the complexities of interrelated and interdependent systems with multiple causal relations. In the end, we are trying to develop students with a global perspective who recognize complex interdependences and consider the future of humanity.

It is time to place the laboratory back in biology instruction. The justifications for laboratory experience far outweigh the excuses for lecture and discussion (Costenson and Lawson, 1987; Mullis and Jenkins, 1988).

An Instructional Model

One of the major problems in biology education is the need for instruction that integrates textbooks, technology, and laboratory experiences. The instructional model proposed here is based on a constructivist approach and has five phases: engagement, exploration, explanation, elaboration, and evaluation. The model includes structural elements in common with the original learning cycle used in the Science Curriculum Improvement Study (SCIS) program (Atkin and Karplus, 1962) and later discussions and research on the SCIS model (Renner, 1986; Lawson, 1988).

The five phases may be summarized as follows:

Engagement

This phase of the model initiates the learning task. The activity should (1) make connections between past and present learning experiences and (2) anticipate activities and focus students' thinking on the learning outcomes of current activities. The student should become mentally engaged in the concept, process, or skill to be explored.

Exploration

This phase of the model provides students with a common base of experience within which they identify and develop current concepts, processes, and skills. During this phase, students actively explore their environment or manipulate materials.

Explanation

This phase of the model focuses students' attention on a particular aspect of their engagement and exploration experiences and provides opportunities for them to verbalize their conceptual understanding or demonstrate their skills or behaviors. This phase also provides opportunities for teachers to introduce a formal label or definition for a concept, process, skill, or behavior.

Elaboration

This phase of the model challenges and extends students' conceptual understanding and allows further opportunity for students to practice desired skills and behaviors. Through new experiences, the students develop deeper and broader understanding, more information, and adequate skills.

Evaluation

This phase of the model encourages students to assess their understanding and abilities and provides opportunities for teachers to evaluate student progress toward achieving the educational objectives.

REFERENCES

AAAS (American Association for the Advancement of Science). 1985. Science Books and Films, 20(5).

Apple, M. 1985. Making knowledge legitimate: Power, profit, and the textbook. In Current Thought on Curriculum. Alexandria, Va.: Association for Supervision and Curriculum Development.

Armbruster, B. 1985. Readability formulations may be dangerous to your textbooks. Educ. Leader. 42(7):18-20.

Atkin, M., and R. Karplus. 1962. Discovery or invention. Sci. Teach. 29: 45-51.

Bybee, R. 1984. Human ecology: A perspective for biology education. Monograph Series II. Reston, Va.: National Association of Biology Teachers.

Bybee, R. 1987. Human ecology and teaching. New trends in biology teaching. UNESCO 5:145-155.

Bybee, R., N. Harms, B. Ward, and R. Yager. 1980. Science, society, and science education. Sci. Educ. 64:377-395.

Bybee, R., P. Hurd, J. Kahle, and R. Yager. 1981. Human ecology: An approach to the science laboratory. Amer. Biol. Teach. 43:304-311.

Carter, J. 1987. Who determines textbook content? J. Coll. Sci. Teach. 16: 425, 464-468.

Commission on Work, Family, and Citizenship. 1988. The Forgotten Half: Non-College Youth in America. Washington, D.C.: William T. Grant Foundation. (ERIC Document Reproduction Service No. ED 290 822)

Costenson, K., and A. Lawson. 1987. Why isn't inquiry used in more classrooms? Amer. Biol. Teach. 48:150-158.

Ehrlich, P. 1985. Human ecology for introductory biology courses: An overview. Amer. Zool. 24:379-394.

Ellis, J. 1984. A rationale for using computers in science education. Amer. Biol. Teach. 64:200-206.

Gould, S. J. January 1988. The case of the creeping fox terrier clone. Or why Henry Fairfield Osborn's ghost continues to reappear in our high schools. Nat. Hist. 19-24.

Hodgkinson, H. 1985. All One System. Washington, D.C.: Institute for Educational Leadership, Inc.

Kaehler, C. 1988. HyperCard power: Techniques and scripts. Menlo Park, Calif.: Addison Wesley Publishing Company.

Koshland, D. 1988. For whom the bell tolls. Science 241:1405.

Lawson, A. 1988. A better way to teach biology. Amer. Biol. Teach. 50: 266-278.

McInerney, J. 1986. Biology textbooks. Whose business? Amer. Biol. Teach. 48:396-400.

Moore, J. 1985. Science as a way of knowing. Human ecology. Amer. Zool. 25:483-637.

Moyer, W., and W. Mayer. 1985. A consumer's guide to biology textbooks. Washington, D.C.: People for the American Way.

Mullis, I., and L. Jenkins. 1988. The Science Report Card: Elements of Risk and Recovery. Princeton, N.J.: Educational Testing Service.

Muther, C. 1987. What do we teach, and when do we teach it? Educ. Leader. 23:77-80.

Renner, J. 1986. The sequencing of learning cycle activities in high school chemistry. J. Res. Sci. Teach. 23:121-143.

Rosenthal, D. 1984. Social issues in high school biology textbooks: 1963-1983. J. Res. Sci. Teach. 21:819-831.

Rosenthal, D., and R. W. Bybee. 1987. Emergence of the biology curriculum: A science of life or a science of living? pp. 123-144. In T. Popkowitz, Ed. The Formation of School Subjects. New York: Falmer Press.

Rosenthal, D., and R. W. Bybee. 1988. High school biology: The early years. Amer. Biol. Teach. 50:345-347.

Weiss, I. 1978. Report of the 1977 National Survey of Science, Mathematics, and Social Studies Education. Washington, D.C.: U.S. Government Printing Office.

Weiss, I. 1987. Report of the 1985-86 National Survey of Science and Mathematics Education. Research Triangle Park, N.C.: Research Triangle Institute.

20
A New Kind of Museum of Natural History as an Instrument of Informal High-School Education in Biology

E. KAY DAVIS

Traditional museums of natural history have long played a public role in the informal education of high-school students of biology. This important function of museums has been shared with numerous other types of institutions, such as zoos, botanical gardens, aquariums, national park projects, and public television production facilities. Doubtless we can agree that all these institutions have made significant contributions to high-school education in biology, both by generally stimulating students' interest in biological science and often by permitting a measure of on-the-scene study and participation in the biological world.

In today's world, however, young people of high-school age have reached new levels of sophistication with respect to what can attract and hold their attention. Educational attractions must now vie with theme parks, with elegant electronics, and with an endless variety of film and television entertainment for the attention of young minds. In order to compete successfully for the time and attention of young people of today and of tomorrow, the institutions of informal education, it seems, are necessarily compelled to rethink, revise, and revitalize their programs.

Yesterday's museums of natural history, for example, were conceived essentially to house and display collections. And at the turn of the last

E. Kay Davis serves as executive director of Fernbank, Inc., in Atlanta and directs a $35 million project to build a museum of natural history. She has a B.A. in biology, an M.A. in science education, and a Ph.D. in administration.

century, that concept made sense: the early museums of natural history allowed the public to see reproductions of animal and plant life that were quite exotic and completely unavailable to most of the public through other media. Today's museums—and especially tomorrow's—cannot, I contend, be based primarily on collections of artifacts. Instead, they must be founded on the concept of delivering information and on offering that information packaged in the most effective and attractive ways.

The Fernbank Museum of Natural History, a new $35 million project just under way in Atlanta, has been conceived expressly to meet these new requirements. Intended to be a museum for the coming century, this project is being designed from the ground up to be a formidable instrument of public education. It is designed with an interesting, definite, coherent story to tell—and everybody loves a good story. Furthermore, it is designed to tell that story with state-of-the-art exhibition techniques and with the flair that has come to be expected of modern entertainment.

To illustrate how we at Fernbank are suggesting that museums of tomorrow may serve the needs of informal high-school education in biological sciences, let me outline briefly some of the major thrusts of this specific museum.

First, the Fernbank Museum is oriented around a central theme exhibit entitled "A Walk Through Time in Georgia." A story line of the 30,000-square-foot exhibit encompasses nothing less than the natural history of the Earth from the "Big Bang" to the present—and even into the future.

Because Georgia happens to enjoy such an unusually varied array of environments—mountains, plateaus, coastal plains, swamps, marshes, and off-shore islands—it is feasible to consider the natural history of the Earth by focusing on Georgia as a microcosm of the Earth. In this way, the museum visitors not only are acquainted with the story of the Earth's natural history, but are specifically acquainted with their immediate environment and how it got to be that way.

In the case of schoolchildren, this format serves not only to acquaint them with the natural history of the Earth and of their immediate environment, but also to engender in them an appreciation of the biological, geological, and physical worlds around them and how they got to be that way—worlds that they can explore and examine in the course of their daily lives.

The "Walk Through Time in Georgia" does not, of course, consider only the biological aspects of the Earth's natural history. It is important that museum visitor recognize, for example, the intricate relationships between the geological history of the Earth or of a specific region and the biological development of life there.

To that end, the story of natural history is presented so that the biological perspectives of the story mesh and blend in with the broader story that includes astronomy, paleontology, geology, archaeology, anthropology, and the physical sciences. Thus, students of biology who visit the museum are given a broad vision of how their formal studies of life on Earth are consistent and interwoven with what is known from other disciplines that they are studying or will study.

Part of the "Walk Through Time in Georgia" concentrates on several geophysical regions of present-day Georgia. In the exhibits that make up this section, the museum visitor is urged to find evidence in today's landscapes that corroborate the natural history that has been elaborated in the chronological sequences. Fossils, rocks, minerals, and geological strata, for example, are made available to be "discovered," providing "clues" to the story that scientists have pieced together.

Another feature found throughout the theme exhibit is a liberal sprinkling of exhibit subsets that pose the question "How do we know?" and then help in answering it. For instance, the purported ages of rocks are supported by small exhibits that demonstrate and explain dating methods using radioactive decay of long-lived isotopes or, for more recent artifacts of native American cultures, the techniques of radiocarbon dating. Thus, the theme exhibit not only presents our present understanding of natural history, but shows why and to what extent we are confident of our knowledge.

Perhaps more important, visitors are encouraged to discover for themselves many of the important relationships between today's built world and its natural history.

The theme exhibit at Fernbank extends the traditional role of museums of natural history by proceeding beyond the present into the future—and not showing the visitor just another "Buck Rogers" vision of how the world might be some time hence. Indeed, the first part of this section is a presentation called "The History of the Future," in which visitors are reminded that, although humans find themselves compelled to contemplate the future, we have never been especially accurate at prognostication. The visitor is asked to consider what has been learned in the museum about the development of humankind in the context of natural history and to focus that new knowledge on how it may help us in making more intelligent decisions and choices for the future. This major section of the theme exhibit, entitled "The City and the Future," assumes that humankind represents a pinnacle of natural history and that the archetypal human habitat, the city, is a reasonable setting in which to celebrate human achievement.

Taking the position that human achievements—including technological, cultural, and even artistic achievements—are part of natural history is far from traditional in the museum world. We feel that, with this point of view, modern museums of natural history may appropriately include components of technology and art—an innovation that we hope will broaden the scope of museums of natural history and increase their attractiveness to the public.

In "The City and the Future," museum visitors are acquainted with modern decision-making aids, such as computer simulations of complex systems. While learning about and interacting with a computer simulation of an urban complex, visitors are introduced to several important concepts. First, they are made aware of the fundamental lesson that the human mind cannot keep track of all the intricacies of a truly complex system and consequently that humans are prone to make well-intentioned, intuitive decisions that often turn out to be counterproductive.

Another important lesson that is emphasized by the computer simulation is that no decision for a very complex system like a city comes about without costs, either in resources or in the quality of life; thus, we are usually faced not with answers, but with tradeoff choices between alternative scenarios. It is hoped that the museum experience will provide our visitors a clearer understanding of the environment—of what affects it and how.

It is further hoped that the museum experience will help our visitors to participate more intelligently in the decision-making processes of which we are all a part. In other words, a museum of natural history may, as part of its informal educational functions, serve a public role in fostering more responsible citizenship.

Apart from the central theme exhibit of the Fernbank Museum, there are further components of the museum that serve exciting roles in informal education. For very young children, there is a Discovery Room, a large area shaped like Georgia. This section both foreshadows and echoes many of the themes that are presented in the "Walk Through Time in Georgia." In the exhibit, small children are provided with "ranger packs" and permitted to slide down Georgia's rivers, hunt fossils and minerals in the mountains, and splash around the swamps and seashores of their state.

There are additional major areas for older children, too. Fostering the natural inquisitiveness and inclination toward hands-on participation of high-school students is a role that museums may undertake effectively if the students can be allowed to participate in personal study and research projects that have been generally unavailable to them in museums of the past. At the Fernbank Museum, a spacious area called the Naturalists' Center is provided to engender participation. The students—adult amateurs, too—are encouraged to use this modern laboratory-library to conduct their own research.

They may bring their own specimens—animals, plants, rocks, insects, fossils, minerals, whatever—for identification, investigation, and study. The amateurs are provided the use of modern, sophisticated, technical instrumentation, available under the supervision of trained staff when required.

The amateurs are also provided with a research library, again with appropriate personnel to provide assistance in finding out more and more information about their own interests. The Naturalists' Center is an attraction that is expected to bring youngsters back to the museum again and again, making the museum a comfortable and productive part of their lives and nurturing their own inclinations toward research.

Considerable space in the Fernbank Museum is reserved for temporary exhibits as they become available for display. The first scheduled exhibit is, for example, the Smithsonian's "Tropical Rainforests: A Disappearing Treasure." Traveling exhibits of this magnitude have heretofore been virtually unavailable to us in the Southeast, because of lack of sufficient exhibition space.

Another attraction, intended to give visitors yet more reasons to make repeated visits to the museum, is the IMAX movie theater, which incorporates a 3-story-tall cinema screen. In this setting, the viewer is engulfed by the projected scenes and feels a sensation of having been thrust into the midst of the action. The professionally produced IMAX films are of exceptional quality, are to be changed on a regular schedule, and are based on appropriate topics for a museum of natural history, ranging from astronautical adventures to thrilling undersea explorations.

Because public education is the primary purpose of a museum of natural history, Fernbank proposes to incorporate an intensive educational program under the direction of a curator of education. The range of courses that have been organized fall, perhaps, into a category more like "semiformal" education. Aside from the customary lectures and seminars that are traditional in museums, the Fernbank museum has organized several sequences of courses and minicourses for the public, for high-school students, and for elementary-school and high-school teachers.

By coordinating the museum's course offerings with the area school systems, it will be possible to have classes of high-school students of biology and earth sciences bussed to the museum for minicourses of relatively short duration. Other, more extended courses, lasting as long as 12 weeks, have been devised for specially selected students. In these courses, the students will be housed dormitory-style in buildings that are on Fernbank property adjacent to the museum. These courses, already designed, are based on the subject material of the theme exhibit of the museum and will feature field trips to all the major geophysical regions of Georgia at appropriate points in the curriculum.

The Fernbank Museum of Natural History expects to achieve considerable leverage in its educational goals by offering specialized courses, again already designed, for both elementary-school and high-school teachers throughout Georgia. These courses, now being approved by the state school system, will afford the teachers graduate credit toward certification advancement. By becoming familiar with the story of natural history as presented in the museum, teachers will be extraordinarily prepared to serve as guides and "talking heads" when their classes visit the museum. It is further hoped that the same teachers will be inspired to serve as statewide "evangelists" of the museum's story and of its informal educational opportunities.

In all aspects of the exhibit program of the new museum, every effort has been made to present the story of natural history using state-of-the-art exhibition techniques. Every effort has been made to make the visitor's museum experience one that is rich in information, beauty, and fun.

Young people will find their visits to this new kind of museum an appealing and repeatedly attractive learning experience that will supplement and amplify their formal education throughout their high-school education and beyond.

21
Messing About in Science: Participation, Not Memorization

CANDACE L. JULYAN

INTRODUCTION

Consider how you first became interested in science. For many people, that interest grew from a curiosity about a particular phenomenon or organism. Satisfying the curiosity, or what David Hawkins (1978) calls "messing about," often resulted in some type of relatively unstructured exploration led by one's own questions, rather than the questions of others. Unfortunately, messing about is not a common practice in many science classrooms today. Students are more likely to be introduced to organisms and phenomena through text-based lectures than by making sense of their own observations. The result of our current science curricula is not only a lack of interest in science as a profession, but also a lack of scientific understanding.

The purpose of this paper is to encourage a consideration of the possibilities of instructional materials that allow students to explore science through projects, rather than texts. While this approach to biology education may differ dramatically from the current practice, I believe that it moves students closer to the experience of science and potentially may offer a deeper understanding of the topics of study. This belief is based

---

Candace L. Julyan is the manager of curriculum and training for the National Geographic Kids Network, a National Science Foundation-funded science curriculum developed at Technical Education Research Centers, Cambridge, Massachusetts. She received a doctorate from Harvard University, and her research has focused on high-school students' understanding of leaves changing color.

on both research and practice: research about how students make sense of science topics and practice as seen in field tests of a new set of instructional materials under development. Although Audrey Champagne has already addressed this area of research, I would like to return to that topic briefly to connect our growing knowledge of students' understanding of science topics with appropriate design for instructional materials.

RESEARCH ON STUDENT LEARNING

In the last decade, educational researchers have introduced new data about how students explain science topics to themselves. Studies based on interviews and teaching situations, rather than tests, encompass a range of topics, such as weight and density (Smith, 1985; Duckworth, 1986), heat and temperature (Smith, 1985), gravity (Stead and Osborne, 1981; Gunstone and White, 1981), light (Stead and Osborne, 1980; Anderson and Smith, 1983), energy (Brook and Driver, 1984), complex systems (Duckworth et al., 1985), and plants (Bell and Brook, 1984; Julyan, 1988). These studies present surprising information about students' preconceptions about how the world works. Although the methods differ, the data suggest that these ideas, referred to by some as "misconceptions," form a basic structure of knowledge that either helps or hinders an individual's ability to make sense of the material presented in the classroom. In addition, these studies suggest that a student's initial beliefs are remarkably resilient and are not erased when a teacher presents new information that might challenge those beliefs. This is an important point to keep in mind: Students do not learn simply by being told.

If correct, how might this supposition affect the work of a science classroom? Osborne and Gilbert (1980) contend that many students merge their erroneous thoughts with the words presented in science class, resulting in continued misconceptions that are now misidentified with a scientific term. These authors believe that by ignoring the ideas that students have before science instruction the teacher may inadvertently continue to support these incorrect notions.

A second concern that is raised by this supposition is that the work of the science classroom cannot rely on lecture- and text-based activities if students are to understand the subject matter. Futhermore, data from several of these studies (Bell and Brook, 1984; Duckworth, 1986; Julyan, 1988) suggest that knowing correct terms does not help students to make sense of their observations. In fact, in many cases, scientific terms may be used to mask confusion. Knowing words does not constitute understanding.

While I am certainly not suggesting a classroom ban on scientific terms, I do suggest that we reconsider their importance. The difference between memorizing vocabulary words related to science topics and understanding various phenomena is considerable. Certainly those of us involved in science and science education realize the difference and are striving for the latter. Science-as-vocabulary requires less effort on the part of both the teacher and the student, but it also provides fewer rewards.

Students, I believe, are also aware of the difference; and if given the choice, they too would strive for understanding. This point was highlighted for me several years ago by a high-school student who was participating in a research study on how students make sense of a complex system (Duckworth et al., 1985). To examine this question, my colleagues and I had devised a number of experiments featuring helium balloons weighted with enough strands of string so that the string dragged on the floor. The students' task was to explain why the balloon's position in the air changed as they did various tasks such as tying knots in the string, cutting the string, putting the balloon on a table, etc.

After several weeks of experimenting, the balloons continued to surprise one student whose frustration with her lack of understanding was often visible. One day, she turned to me demanding to know whether I intended to tell her "the answer." When I asked her to state the question, she seemed momentarily stumped, but then stated that she wanted to know why the balloon behaved in the way that it did in all the various experiments that she had conducted. While certainly willing to answer her question, I stated that I wanted to be clear about what she wanted. Did she want words that explained the phenomenon or did she want to understand it herself. There was a long and poignant silence in the room. Finally, she turned and quietly said that the words were *not* what she wanted; she really wanted to understand.

While most students might not articulate the dilemma as clearly as this student, many share her desire to understand, not just to know the correct words. One way to promote both understanding and the value of scientific terms is to give students an opportunity to mess about in their science classes, to become participants in constructing their knowledge. Inquiry-based activities are certainly not the fastest way to approach science learning. A faster and more "efficient" approach is text-based, lecture-based classes. This approach, which is certainly the predominant approach in science classes today, is not terribly effective, as indicated in the numerous reports about science education and the low enrollment in science classes. Change is definitely needed. Let me turn now to a curriculum that offers this type of change.

A NEW TYPE OF SCIENCE CURRICULUM

While the importance of inquiry in the science classroom is certainly not a new idea, my colleagues and I at Technical Education Research

Centers (TERC) are developing a curriculum that has a new commitment to the notion of inquiry. Because its operating premise about what is possible in a science classroom is unusual, I offer it as an example of an unconventional approach to instructional materials worth noting. The basic premise of this curriculum, the National Geographic Kids Network, is that students can and should *be* scientists, that they can and should converse with real scientists about their work, and that computers can enhance this enterprise. Students, therefore, conduct experiments, analyze data, and share their results with student colleagues using a simple computer-based telecommunication network. This collection and making sense of data gives these students an opportunity to experience the excitement of science that scientists feel.

We are just beginning the third of 4 years of this National Science Foundation-funded project, which represents a partnership between TERC and the National Geographic Society (NGS). TERC is developing the curriculum and software for the project; NGS will publish and distribute the final product. The full curriculum, designed for upper elementary school, grades 4 through 6, will consist of a number of 6-week units, which are intended to supplement the regular curriculum.

Each unit focuses on an environmentally oriented topic, such as acid rain, land use, or water quality. Students begin their study by examining the topic in the context of their local community and then exploring it within the larger national picture. Each unit involves the collection of some type of data (survey or measurement), sharing those data through a telecommunication network with other classes collecting the same data, and finally making sense of those data. These data are examined by both student colleagues and a unit scientist, a professional with expertise in the unit topic.

The computer gives students a number of tools to help with the data collection and analysis: a word processor, a record-keeping data section, a graphing utility, a complete telecommunication package, and map software with data overlay. This software component of the project is both simple and powerful.

The Kids Network is more, however, than powerful software; it is a careful weaving of classroom activities and software tools. The following outline of the acid-rain unit illustrates the connection that exists between these two components of the project.

Students begin this unit by learning how to use pH paper and how to build a rain collector. With this knowledge and equipment, they record the pH reading for each rainfall for the next several weeks. While waiting for the rain, they continue to explore pH through experiments that look at the effect of solutions of different pH on the growth of seeds and on a variety of nonliving objects. They also keep a weekly log of odometer readings from the family car and roughly calculate the amount of nitrogen oxide that they have introduced into the atmosphere as a family and as a class. Through letters, classes can share the findings from all these activities with others in the network. These letters are sent, not to the thousands of schools in the network, but to their "cluster," a small group of 10-12 geographically dispersed classrooms.

After several weeks, during which time one hopes there has been some rain, classes send their rain data to the network using the data-entry feature of the software. Those data are collected and sent back within a week.

Students receive the network data, as well as a letter from the unit scientist. The data may come back as a data-entry form or as a map file with color-coded data from each site displayed. The letter from the unit scientist, in this case John Miller of the National Oceanic and Atmospheric Administration (NOAA), helps students to put their data into the context of other data collected on acid rain. Next, students are asked to compare their readings with those of their fellow cluster classes and to examine patterns in the network data. Again, a flurry of letters to fellow scientists takes place, asking for clarification about surprises or validation of theories.

The unit ends with a look at the social significance of the data. Students examine two very different positions on what to do with the information gathered. Each position, seemingly written by a student-colleague on the network, addresses the impact of decisions about acid rain—e.g., a loss of jobs for factory-working parents or serious reduction of fish in a local pond. Students are encouraged to discuss these positions, take a class vote for no action without further research or for both immediate action and further research, and send the results to the network.

WHAT MIGHT THIS CURRICULUM DESIGN SUGGEST?

The Kids Network has generated an enthusiastic response on the part of both teachers and students. As one teacher explained, it "is an awesome concept, a truly revolutionary idea for education at a time when it is so badly needed." We have completed the design of two units and have tested these preliminary materials in 200 classrooms across the United States and in a handful of foreign sites, including Canada, Argentina, Hong Kong, and Israel. Data from our formal evaluation, which included both observations and questionnaires, and the flood of unsolicited narratives received from teachers through phone calls and letters suggest that teachers and students found this curriculum to be a lively and appealing way to approach science.

Despite the elementary focus of these materials, there are a number of ideas or issues that are applicable to the high-school curriculum. I propose these ideas in the form of questions. While all the questions represent ideas around which the Kids Network has been designed, they also provide a framework for examining aspects of this project that may provide new ideas for those interested in high-school biology curricula. After briefly noting examples of the ways in which our curriculum addresses each question, I will suggest areas of consideration that these questions raise for instructional materials.

Can Data Collection and Analysis Provide an Effective Backbone Around Which to Study Science?

Teachers and evaluators reported numerous examples of scientific thinking generated by the simple act of collecting and making sense of data. For example, in the "hello" unit, students were collecting data about the kinds of pets that each student cared for. Many classes found that defining their terms, in this case what constituted a pet, was quite complicated. In one class, the discussion revolved around a debate as to whether an ant farm should be considered a pet. The class finally decided that a pet had to share the same environment with its owner. Therefore, they decided that an ant farm was not a pet, as it was in a sealed container; but fish were pets, as they lived in water from a faucet. Although you may argue with their decisions, I think that you can see that the students did exhibit scientific thinking as they tried to resolve their question.

Another example of the type of scientific understanding that data collection generated came from a fourth-grade class. These students realized that, although their data and those of their cluster school were both about pets, they could not compare data. One class had recorded the number of students who owned each kind of pet; the other had recorded the number of each kind of pet. The class learned a great deal about the importance of comparable data, a concept that many high-school students might find difficult.

Can Inquiry-Based Instruction Help a Larger Proportion of Students to Feel Confident About Their Ability to Understand Science?

This question was addressed in several ways. Many teachers reported that students were motivated and eager to participate in the curriculum activities, even, or perhaps especially, students who rarely participated in science class. One teacher told the story of a "low-ability" student who gained enormous credibility in class when he proposed his idea about the wide discrepancy between the over 120 pets owned by his Auburn, Maine, class and the fewer than 25 pets owned by a cluster class in New Orleans. While other students had explained the difference as something related to the weather or the availability of pets in Louisiana stores, this student thought that perhaps the school was located in an area surrounded by government housing. He explained that pets were not allowed in this type of housing. This simple piece of information from a fellow student, not the teacher or a textbook, helped students to make sense of the data and generated a letter to find out whether this student's theory was correct. As you might imagine, his idea also increased the student's credibility with his peers, his teacher, and perhaps himself.

We received other reports of student self-esteem that are of a very different nature. Teachers reported that students felt a real sense of ownership about their work. The most extreme examples came from two teachers who reported that their classes, both filled with "average" and "below-average" students, protested on learning that their teacher was planning to present the Kids Network curriculum at professional conferences. The students argued that it was *their* work and that *they* should be the ones presenting. In one case, the teacher took the students with him; in the other, the students made an acetate filmstrip and accompanying audio tape that the teacher used in her presentation.

Can Technology Enhance Inquiry in the Science Classroom?

Bob Tinker (1988), the project's director, contends that "technology can bridge the gap between the conduct of science and the teaching of science." The Kids Network provides this bridge in several ways. First, by using microcomputers for computation, students are able to manipulate their findings in more sophisticated ways than their computational skills might have permitted otherwise. Second, with the power of telecommunication, students are able to share data and ideas with others from all over the country. This extension of the classroom provides a powerful motivation for many students. Lastly, telecommunication offers a unique and manageable opportunity for scientists to communicate with science classrooms. The technology expands these classrooms by eliminating the limitations of both time and distance that would otherwise restrict this type of communication.

Can the Work of a Science Classroom Generate Community Interest?

Kids Network classrooms were filled with visitors interested in the students' work. The list included principals, other teachers, superintendents, school-board members, parents, university professors and their students, and reporters from newspapers and television. These visitors were interested in what the students were doing, in terms of both the telecommunication activities with other schools and the data they were collecting. Adults were as interested in the data results as the students, particularly when the data were about a larger environmental concern, such as acid rain. This type of interest provided a motivation for both teachers and students, who appreciated having an audience for their work.

Can the Work of a Science Classroom be of Interest to the Scientific Community?

A portion of my work on the curriculum component of this project is to find experiments appropriate for elementary-school students and useful to research scientists. This search has been one of the most challenging and delightful aspects of my job. The scientific community has been both supportive and enthusiastic. Many scientists have given hours of their time exploring the types of research that students might do that would offer valuable data to existing studies.

John Miller, the deputy director of the acid-deposition unit of NOAA and our unit scientist for the acid-rain unit, is an excellent example of the type of support and enthusiasm that we have found. Dr. Miller corresponded throughout the unit with his elementary-school "colleagues" and has proposed that the Kids Network data be included as an appendix to the NOAA 1988 report on acid rain. In addition, he regularly discussed this project with his colleagues around the world. Perhaps the most surprising example of the serious interest that scientists have in student-collected data took place at the annual meeting of a United Nations steering committee concerned with the long-range transmission of air pollutants. As the United States representative to this group, Dr. Miller presented an overview of the various networks in North America that are concerned with acid rain. Within this context, he introduced his colleagues to the Kids Network project and was surprised to find that they expressed considerable interest in this network, wanting to know how students made the measurements and whether the data would be available for comparison with NOAA data. We have found that this level of interest has not been unusual. Scientists are interested in student measurements, particularly when the data cover a wide geographic area.

CONCLUSION

The first reports about the effectiveness of the Kids Network indicate that it generates considerable enthusiasm in both students and teachers and enlivens the science classroom. In times filled with grim reports about science education, this curriculum has sparked considerable hope. Our experience with elementary-school students has shown that the approach has great promise to change students' perception of science and of their abilities in the subject. It offers them a collective opportunity for messing about in science.

This same type of exploration is certainly possible for high-school classrooms and is worthy of further consideration. Instructional materials that center around the collection of data not only give students a more accurate picture of the scientific enterprise, but they also give students an opportunity to examine their personal ideas about the topic under study. In addition, if this type of investigation is linked to real environmental concerns, the students' work yields valuable community information and is not just an empty school assignment.

Telecommunication is a new and exciting tool for the science classroom. It provides an easy way to connect the work of various classrooms and to help students to understand the larger scientific picture into which their experiments fit. One difficulty that many existing telecommunication networks report is that these connections die if there is no reason to communicate (Carl Berger, University of Michigan, personal communication). Creating a network that revolves around the collection and analysis of data provides an important and engaging topic for conversation.

While the telecommunication activities of the Kids Network curriculum are seemingly elaborate, classroom exchanges can be fairly simple and do not require either fancy software or equipment. All that is required is a worthwhile set of data to collect and an interest in sharing those data among classes as close together as the same town or as far apart as different coasts. Science teachers and curriculum developers may be surprised to discover the number of groups that would welcome this sort of low-cost data collection and would help with the necessary logistics.

Finally, I would like to note that, while the technology of this curriculum suggests an exciting new resource for the classroom, the Kids Network suggests a very simple change for classrooms. This change, while enhanced by technology, does not require it. Messing about in science does not require fancy equipment, just an appreciation of the value in making sense of the many mysteries of life. Certainly that should be an important value for biology education. One high-school student involved in examining trees was pondering the difference between his biology class and his work in an inquiry-based study. "I don't know why we read about trees in science class. It seems stupid not to come outside and really study 'em. Don't ya think?" (Julyan, 1988).

I obviously agree, but hope that I have provoked the consideration of new possibilities that might exist in future high-school biology curricula. These possibilities could include materials that center around students collecting original data, students sharing those data with interested student colleagues, and teachers and scientists working together to help students to make sense of their findings. Materials that have these various foci would certainly differ from those in classrooms today—differences that might alter both biology education and perhaps biology itself. I am hopeful about the future of biology education, particularly if it encompasses these simple ideas. I believe that students and teachers are ready for a change.

REFERENCES

Anderson, C. W., and E. L. Smith. 1983. Childrens' Conception of Light and Color: Developing the Concept of Unseen Rays. Paper presented at the annual meeting of the American Educational Research Association, Montreal.

Bell, B. F., and A. Brook. 1984. Aspects of Secondary Students' Understanding of Plant Nutrition. Leeds, England: Center for Studies in Science and Mathematics Education, University of Leeds.

Brook, A., and R. Driver. 1984. Aspects of Secondary Students' Understanding of Energy. Leeds, England: Center for Studies in Science and Mathematics Education, University of Leeds.

Duckworth, E. 1986. Inventing Density. Grand Forks, N.D.: University of North Dakota Press.

Duckworth, E., C. Julyan, and T. Rowe. 1985. A Study on Equilibrium: A Final Report. Cambridge, Mass.: Educational Technology Center.

Gunstone, R. F., and A. White. 1981. Understanding of gravity. Sci. Educ. 65(3):291-299.

Hawkins, D. 1978. Critical barriers to science learning. Outlook 29:3-23.

Julyan, C. 1988. Understanding Trees: Five Case Studies. Unpublished doctoral dissertation. Harvard University.

Osborne, R., and J. K. Gilbert. 1980. A method for investigation of concept understanding in science. Eur. J. Sci. Educ. 2:311-321.

Smith, C. 1985. Student Conceptions of Heat and Temperature. Cambridge, Mass.: Educational Technology Center.

Stead, K. E., and R. Osborne. 1980. Exploring science students' concepts of light. Austral. Sci. Teach. J. 26(3):51-57.

Stead, K. E., and R. Osborne. 1981. What is gravity? Some children's ideas. New Zealand Sci. Teach. 30:5-12.

Tinker, R. F. 1988. Telecommunication and Science Education. Talk delivered at the AAAS conference, February 1988.

PART V

Teacher Preparation

22

Biology Teacher Education: Panacea or Pitfall

JANE BUTLER KAHLE

Recent reports of international and national assessments of science education (IEA, 1988; Humrich, 1988; Rothman, 1988) tell the same old tale: most of our students are inadequately prepared to do science, to solve science-related problems, or to resolve science-related social issues. For example, results of the second International Association for the Evaluation of Educational Achievement (IEA) science study show that, compared with the science achievement levels of their peers in other countries, our fifth-graders rank eighth of 15, our ninth-graders rank fifteenth of 16, and our advanced students who have had *2* years of biology rank last in a field of 14. Increasingly, we are educating citizens who are scientifically illiterate. Education in biology, among the sciences, is in a key and pivotal position to alleviate the current situation—key, because it is the only science studied by the vast majority of our students, and pivotal, because study of biology may provide the motivation, stimulation, skills, and interest that encourage a child to elect optional science courses, such as those in chemistry and physics. Therefore, biology teacher education may be the panacea or the pitfall as we attempt to reform science education in our schools. My remarks are intended to provide an overview of the current situation so that we can formulate directions for change.

Jane Butler Kahle is Condit Professor of Science Education at Miami University and a former president of the National Association of Biology Teachers. She has published numerous articles on teaching secondary-school science and on women and minorities in science education.

The education of biology teachers is usually a two-part process: preservice (or undergraduate) education and in-service (or postgraduate) education. In the United States and most other countries, the preservice education of biology teachers consists of two components: general undergraduate education and specific professional training. The undergraduate biology and related science education of prospective American teachers varies greatly, depending on the type of institution they attend. For example, at most liberal arts colleges, prospective biology teachers complete the same courses as do all biology majors. In colleges specializing in teacher training, the content courses may be special ones that are directed toward specific teacher licensing requirements. The type of biology background received at large universities varies according to the structure of the university; that is, biology courses for teachers may be taught under the aegis of the college of education or of the college of arts and sciences. Both the type of course and the level of competition may be radically different, depending on the organization of the teacher education program within the university. Furthermore, biology teachers may be certified by religious colleges and universities that do not recognize or teach evolution as the unifying theme of biology.

While the median required number of hours in biology for teacher certification in the United States is 24, 21% of biology classes are taught by teachers who have had less than 18 hours in undergraduate biology classes (Helgeson et al., 1977). Generally, undergraduate courses taken by preservice biology teachers are the same courses taken by students preparing for professional or graduate schools. Large, impersonal lecture courses with structured laboratories are the most common format, and prospective teachers have few opportunities to participate in long-range laboratory inquiries, to lead fruitful discussions, or to ask and respond to penetrating questions. The infrequent use of creative inquiry teaching in biology classrooms may be related to the fact that teachers rarely experience it in their college preparation. Currently, the education of biology teachers is considered by many to be a pitfall.

About 4 years ago, many professional societies and accreditation groups examined the teacher education process in general and recommended sweeping changes. Generally, the reform movement has focused on higher standards of admission to teacher education programs. Although the situation is not as drastic in biology, generally prospective teachers have verbal and quantitative Scholastic Aptitude Test scores 32 and 48 points, respectively, below the scores of students choosing other careers. And the bright students who indicate an interest in teaching are routinely discouraged from it by professors of biology. In addition, an academic major in biology and a subsequent professional internship are recommended. Teacher assessment, including success on the national teacher examination, is another focus of the reform movement.

One model, formulated from the proposals of various reform groups, suggests that prospective teachers complete a baccalaureate degree in the biological sciences and then choose between two routes: a fifth year of professional courses, including an internship, which leads to certification, but not to a graduate degree; or 2 years of further study and internship, which lead to certification and a master's degree. One purpose of this model is to develop a training program for professionals. A second purpose is to develop the commitment needed to keep professional teachers in the classroom; more highly qualified teachers are more likely than others to leave the workforce (Darling-Hammond, 1984). Of great concern is the fact that 21% of minority-group teachers, compared with 12% of non-minority-group teachers, who state that they are very satisfied with their careers plan to leave teaching (Daniels, 1988).

One of the problems with the current reform movement in teacher education is that the institutions that educate the majority of teachers are not part of it. Perhaps, because of their exclusion or lack of involvement, traditional teacher education institutions have questioned the recent reforms. Members of the Teacher Education Council of State Colleges and Universities (TESCU) have issued a response to the initiatives of the Holmes Group (Backman, 1986). Briefly, they propose retaining the education major, retaining the standard 4-year program, requiring the same certification standards for all teachers, eliminating the establishment of national standards for teacher competence, and strengthening the role of the school principal. Many of TESCU's proposals counter, or at least dilute, the reforms promulgated elsewhere. Yet, TESCU institutions educate the vast majority of our teachers. For example, in 1985 TESCU institutions awarded an average of 360 baccalaureate teaching degrees per institution (AACTE, 1986), whereas the research-oriented institutions in the Holmes Group awarded an average of 194 baccalaureate teaching degrees (Olson, 1986). Any effective reform must extend to a variety of institutions or have the power to reduce the number and type of institutions that prepare biology teachers.

Concomitant with the reform movement has been the proliferation of alternative certification programs. This year, about 2,000 teachers will be trained by alternative programs by which people with baccalaureate degrees earn certification on the job. Twenty-three states have changed some of their requirements, and 16 others are considering proposals (Warren, 1988). It is too soon to evaluate the effectiveness of such programs. Similar to the alternative certification option is the use of emergency certification by which teachers are reassigned to other areas—for example, agriculture teachers to biology.

The movement to make teacher education a panacea, rather than a pitfall, has two thrusts. One is to move teaching toward the status of a profession, with rigorous entry requirements, supervised instruction, autonomous performance, peer-defined standards of practice, and increased responsibility with increased competence (Darling-Hammond, 1984). The other views teaching as an art, which any educated person can practice. We need to provide perspective and suggestions as we strive for ways to make biology teacher education a model for science education.

REFERENCES

AACTE (American Association of Colleges for Teacher Education). 1986. AACTE Directory 1986-1987. Washington, D.C.: AACTE.

Backman, C. 1986. Positions on Current Issues in Teacher Education. Washington, D.C.: Teacher Education Council of State Colleges and Universities.

Daniels, L. 1988. More minority teachers may quit. New York Times, October 7:B12.

Darling-Hammond, L. 1984. Beyond the Commission Reports. Series R-3177RC. Santa Monica, Calif.: The Rand Corporation.

Helgeson, S. L., P. E. Blooser, and R. W. Howe, Eds. 1977. The Status of Precollege Science, Mathematics, and Social Science Education: 1955-1975. Volumes I, II, and III. SE 78-73. Washington, D.C.: U. S. Government Printing Office.

Humrich, I. 1988. Sex Differences in the Second IEA Science Study: U.S. Results in an International Context. Paper presented at the National Association for Research in Science Teaching, Lake of the Ozarks, April 10-13.

IEA (International Association for the Evaluation of Educational Achievement). 1988. Science Achievement in Seventeen Countries: A Preliminary Report. New York: Pergamon Press.

Olson, L. 1986. Indiana University's status in Holmes group uncertain. Educ. Week November 26.

Rothman, R. C. 1988. Science: Achievement levels on test "distressingly low." Educ. Week September 28.

Warren, W. J. 1988. Alternative certificates: New paths to teaching. New York Times, September 28:B10.

23

Professional Teachers for High-School Biology

ALPHONSE BUCCINO

A MATTER OF PERSPECTIVE

Programs for the education of teachers must be designed with a vision of the results they are to achieve. I propose that our model program aspire to a vision of a master teacher, which is a status achieved as a result of a significant developmental process that extends over some period and encompasses preservice preparation and subsequent professional practice coupled with continuing professional development involving formal study.

This perspective suggests that we must be concerned not only with what teachers should know and be able to do, but also with the context in which they must practice their profession. Thus, as we move forward with the design and improvement of our programs for the preparation and continuing professional development of teachers, we should note that all current proposals, including those of the Holmes Group (1986) and the Task Force on Teaching as a Profession (1986), address twin goals: to reform teacher education and to reform the teaching profession. Inclusion of the latter goal indicates recognition that the quality of teaching in our schools depends on several factors in addition to the intrinsic quality of teacher preparation programs in our universities.

Alphonse Buccino is dean of the University of Georgia's College of Education and professor of mathematics education. Before being named education dean, in April 1984, Dr. Buccino served in several professional positions in science and engineering education at the National Science Foundation. He received undergraduate and doctoral degrees from the University of Chicago.

This paper further extends the twin goals by separating each into two others: teacher preparation programs comprise content and pedagogy, while the profession is affected by certification and influences on professional identity. This paper emphasizes two of these four elements: pedagogy and professional identity, especially as the latter is or was affected by the National Science Foundation (NSF) teacher institutes. However, for completeness of the perspective indicated here, remarks are presented about content and certification.

CERTIFICATION

"Teacher certification" is the name given the licensing of teachers. A license to teach is required by each of the 50 states, although the criteria and standards for certification (licensure) may vary from state to state. Because certification in all states requires higher education on the part of candidates, the degree programs of colleges and universities strongly influence teacher certification criteria and standards. However, certification also responds to the perceived supply of and demand for teachers and to hiring practices and needs. School managers (i.e., principals, superintendents, and their designees, who are a powerful force as an informal lobby) want as much flexibility as the managers of any organization. They want less regulation, rather than more.

As a result, one sees a tension between the generalist approach and the specialist approach to teacher preparation and certification. The trend nationally in all states is to certify generalist science teachers—despite the efforts of professional organizations, such as the National Science Teachers Association (NSTA) and the National Association of Biology Teachers (NABT), to develop standards for teachers who will specialize. In Georgia, for example, certifying biology, physics, chemistry, or earth science specialist teachers is possible, but the certification of the generalist science teacher for all four subject areas is not only possible, but preferred by school officials for reasons cited earlier. In fact, the generalist certification is the more prevalent route for science teachers in Georgia and elsewhere. However, there is no degree program in the university preparing the generalist teacher, who is so certified by a special review by the state department of education of his or her course record.

Another aspect of certification that should be mentioned is what I call the "back door." Teaching is the one profession where quantity considerations tend to take precedence over quality. To avoid the phenomenon of the empty classroom, all but two of the 50 states have mechanisms that provide substandard, limited, or emergency licenses to persons lacking the qualifications for full professional certification. Clearly, these back-door devices can lower the quality of persons who actually teach in the schools, independently of the quality of teacher preparation programs in colleges and universities.

TEACHER KNOWLEDGE: CONTENT

Aside from the impact of certification standards on the content knowledge of those actually teaching biology in the nation's classrooms, there are other factors to consider. To begin with, the responsibility for content in the preparation of biology teachers (in terms of both the major in biology and the content courses that are part of general education) lies outside the jurisdiction of the school or college of education, which generally has the primary responsibility for teacher preparation. There is much room for improvement in the colleges of arts and sciences of our universities as regards their part of teacher education. Unintegrated programs in general education, excessively narrow subject-area majors (depth of subject without breadth of subject), and poor teaching role models abound. Moreover, the school or college of education needs to have some impact and oversight, which it may not now have, regarding the content experiences of future teachers.

The content knowledge required for biology teaching is set forth in many places, notably the aforementioned NSTA and NABT standards. As Shulman (1986) points out, the teacher's knowledge must go beyond concepts and facts of a domain to an understanding of the structures of a subject. For example, the biology teacher must understand that there are a variety of ways of organizing the discipline, as is reflected by the yellow, green, and blue versions of the Biological Sciences Curriculum Study texts.

In addition to these considerations, the content base for biology teaching is especially complex, owing to four factors. First is the need to include basic mathematics, chemistry, physics, and earth and environmental sciences in the program. Students majoring in other subjects are less dependent on the study of other disciplines. Second, biology is at the interface between natural science and social science. This requires successful teachers to have sophisticated knowledge of both. Third, there is pressure to include medicine in biology education. Fourth, the controversial and so-called science-technology-society issues are prominent in biology, such as the ethical issues and societal problems associated with such phenomena as in vitro fertilization, genetic engineering, and AIDS.

TEACHER KNOWLEDGE: PEDAGOGY

Pedagogical Knowledge of Teaching

Not long ago, a legislator asked me whether I might provide him with some pointers and techniques for dealing with a class of fourth-graders that he was scheduled to visit. With the help of faculty members in our college of education, we produced a page or so of notes that included such items as the following: fourth-graders have short attention spans, so keep the interactions on a given topic brief; children of this age are beginning to develop a sense of independence, but still have some dependence characteristics, so they like being treated like "big" boys and girls; and they love presents. The legislator followed the advice we gave him, remembering also to bring a little gift for each student. He credited the information we provided with the enormous success he later reported of the visit.

The kind of information we provided the legislator might be called "pedagogical knowledge of teaching"—what Shulman (1986) refers to as "generic principles of classroom organization and management and the like." It is widely agreed that such knowledge is essential for effective teaching and must be included in successful teacher preparation programs. Moreover, this kind of knowledge is commonly recognized as the responsibility of the school, college, or department of education—in contrast with subject-matter content, which falls in the jurisdiction of the arts and sciences components. Without meaning to minimize or diminish the importance of this kind of knowledge on the part of teachers, I do want to emphasize another kind of pedagogical knowledge.

Pedagogical Knowledge of Content

This other kind of pedagogical knowledge focuses on the question: What does one need to know about a subject in order to teach it? Great teachers from the beginning of history have known that a special kind of knowledge is associated with teaching a specific subject. Scientists, especially, often speak of their desire to help students to grasp the power and beauty of science, quite beyond facts and concepts. We owe a lot to Shulman (1986) for successfully and pointedly calling our attention to pedagogical knowledge of content as a fundamentally important element of teaching and teacher preparation:

> for the most regularly taught topics in one's subject area, the most useful forms of representation of those ideas, the most powerful analogies, illustrations, examples, explanations, and demonstrations—in a word, the ways of representing and formulating the subject that make it comprehensible to others. Since there are no single most powerful forms of representation, the teacher must have at hand a veritable armamentarium of alternative forms of representation, some of which derive from research whereas others originate in the wisdom of practice.
>
> Pedagogical content knowledge also includes an understanding of what makes the learning of specific topics easy or difficult: the conceptions and preconceptions that students of different ages and backgrounds bring with them to the learning of those most frequently taught topics and lessons. If these preconceptions are misconceptions, which they so often are, teachers need knowledge of the

strategies most likely to be fruitful in reorganizing the understanding of learners, because those learners are unlikely to appear before them as blank slates.

This kind of knowledge is a special form of content knowledge and is, therefore, subject-specific. Once it is identified, it is immediately clear that pedagogical knowledge of content is quite important for teachers and, consequently, should have a central role in teacher preparation. Unfortunately, the opposite seems to be the case. By "opposite" I mean that, despite the preoccupation with subject matter in current concerns about education, and despite the evident significance of pedagogical knowledge of content for the teaching and learning of subject matter, the emphasis in teacher education is on the generic, and not the subject-specific.

For one thing, it is not clear whether primary responsibility for pedagogical knowledge of content in university teacher education programs lies in the college of arts and sciences or in the school or college of education. Both education and arts and sciences have a role, but the degree of emphasis may differ. On the arts and sciences side, this kind of knowledge is often recognized. The physicist J. Robert Oppenheimer, in his invited address to the American Psychological Association (Oppenheimer, 1956), discussed the role of analogy in the development of knowledge in physics. He described five examples of the use of analogy in atomic physics to illustrate his argument that "analogy is indeed an indispensable tool for science."

If analogy is essential for the creation of new knowledge in science, it surely is essential for the teaching and learning of it. Unfortunately, the use of analogy in this regard is of uneven effectiveness. Glynn et al. (1989) studied the use of analogies in high-school physics textbooks. They found that, while all texts used analogies, the use in some instances was much more effective—in pedagogical terms—than in others.

On the school or college of education side, Shulman (1986) points out that research paradigms on the study of teaching are characterized by the omission of one central aspect of classroom life: the subject matter. In fact, Shulman and his colleagues refer to the absence of focus on subject matter among the various research paradigms for the study of teaching as the "missing paradigm" problem. Thus, a premier issue in teacher education today is subject-specific versus generic pedagogical knowledge of content. For teacher education, the missing research paradigm translates into an imperative to integrate content and pedagogy.

Unfortunately, as indicated earlier, the trend is in the opposite direction. There are two forces I would describe as spurious that are driving schools and colleges of education to the generic at the expense of the subject-specific. First and foremost is size and efficiency. Many teacher education programs are so small that they cannot afford to distinguish science from mathematics, or even distinguish science from the humanities, such as language arts. One response—albeit politically difficult, if not unrealistic—is to terminate programs that lack the requisite critical mass. That would mean termination of about half the teacher education programs currently on the books in American colleges and universities.

But even in large programs that have the wherewithal to specialize, the trend is to the generic. The second spurious force acting on teacher education programs concerns program identity. More and more, schools and colleges of education are suppressing specialization within themselves in the interest of projecting a common and unified image. Faculties are choosing to add "common experiences" for all prospective teachers, rather than several diverse subject-specific or otherwise specialized approaches. More and more, institutions call attention to their respective teacher education programs, emphasizing generic characteristics, while few call attention to programs preparing subject-specific high-school teachers of biology, mathematics, physics, or chemistry.

The Role of Theory

This appeal for special attention to pedagogical knowledge of content in the preparation of teachers challenges us to seek learning theories for guidance. Social learning theory (Bandura, 1973) reminds us, among other things, of the truism that we teach as we were taught. Thus, in a real sense, the preparation of future teachers begins in the early grades. However, there surely is an admonition here for college and university faculty members who teach teachers, to model the sort of teaching behavior they expect their students to exhibit.

Current theory has been strongly influenced by research about misconceptions associated with science learning (alluded to in the Shulman quotation presented above). As I observed earlier (Buccino, 1985), this research calls into question the strong tendency to present science as a fresh new subject, something the students have not really encountered or experienced before. The child is the clean slate on which the new scientific information is to be inscribed, or the empty vessel into which fresh scientific knowledge is poured.

This research on misconceptions lends support to constructivist or modern cognitive theory that questions the accuracy of the clean-slate and empty-vessel metaphors. Contemporary thinking and research about teaching and learning indicate that children *construct* their own understanding; they do not just reflect what is given to them. Moreover, the learner's formulation of understanding is based on a great deal of prior information. A child's cultural and familial environments affect how information transmitted in a classroom is processed in the child's mind. These "environmental" factors affect what is retained by the child, what is pursued further, and what is virtually ignored. Thus, children come to their first and subsequent science classes with surprisingly extensive theories about how the natural world works. These naive theories affect what they perceive to be happening in the classroom and in laboratory experiments. These naive theories are developed as a natural human tendency to come to grips with and find order in a world that, especially to a child, seems incredibly complex. Moreover, they often continue to attach their incorrect and naive understandings to situations even after instruction supposedly provides correct versions (Resnick, 1983).

The misconceptions of students are especially acute in biology, because a child's thinking is animistic. We know that the sun is alive because it gives heat, that a stone is alive because it can move as it rolls down a hill. Today, animism extends to computers as children consider that the device with which they may interact is alive (Turkle, 1984). The animistic thinking is only gradually overcome by knowledge and experience. It is unclear how this affects students in high school, but the teacher surely should be prepared for surprising and incorrect naive understandings on the part of students.

From this perspective, pedagogical knowledge of content is all the more important. Moreover, in science the special role of laboratories and demonstrations makes even greater demands. As though this were not enough of a challenge, ways of treating controversial subjects and science-technology-society issues also make special demands not easily dealt with from a base of subject-matter knowledge alone.

Continued Professional Development

Given these complexities, let us recall the vision of the master teacher with which we began this discussion. It is clear that the initial preparation of teachers must be followed by additional professional development.

The structure of our graduate teacher education programs at the University of Georgia illustrates the foregoing discussion. The University of Georgia's teacher education programs based in the College of Education are unusual for their subject-specific emphasis, in both organizational and program terms. These programs are organized into departments that represent content domains. Specifically included are departments of language education, mathematics education, science education, and social-science education. Each of these departments offers programs for teachers, including programs at two graduate levels, and each has nine to 14 faculty members who are expert in both the content and the pedagogy of their subject areas. This is in direct contrast with the predominant generalist approach to staffing in many teacher education programs. Teacher education students at the University of Georgia do not take general courses, such as curriculum, instruction, or computer applications. Instead, they enroll in programs that are content-specific. On the other hand, teacher education students do not take any content knowledge courses in the College of Education. All these offerings are in the College of Arts and Sciences. The critical mass of faculty and students in each content area greatly facilitates this arrangement and allows it to be cost-effective.

THE TEACHING PROFESSION

Is teaching a profession? Shulman (1986) identifies three kinds of knowledge that a teacher ought to have: propositional knowledge, case knowledge, and strategic knowledge. Propositional knowledge arises from research and experience. Propositional knowledge represents the wisdom of practice and the accumulated lore of teaching. Because this is incomplete, we also need case knowledge to provide details and contexts. Propositional knowledge and case knowledge on the one hand imply and on the other hand can be understood only in a conceptual or theoretical framework.

This is where strategic knowledge comes in—the knowledge a teacher needs to make a judgment on how to proceed when propositional principles conflict or cases appear incompatible. The reliance of teaching on strategic knowledge is what makes it a profession, and not merely a craft (Kilpatrick, 1987). Following Shulman (1986), Kilpatrick argues that a craft may be mastered by learning to follow a set of rules, while a profession requires that indeterminate rules be applied to particular cases. The professional practitioner is someone who has developed an awareness of the reasons for making professional decisions. Thus, a teacher is not merely a practitioner, but a thinking practitioner. From this perspective, a teacher does not merely respond automatically to a teaching situation, but makes reasoned judgment through a self-conscious conversation with the teaching situation. Clearly, this view of the teacher as thinking practitioner calls for an end to the separation of research and practice in teaching. Accordingly, teachers need to be helped to redefine the role of research in their work and to recognize that research not only is applied to practice, but also grows out of it.

But in terms of current practice, is teaching a profession? Despite the foregoing argument that teaching is (or should be) a profession, rather than a craft, many observers have noted that, like many occupations, teaching has become seriously deprofessionalized.

The following conditions appear to be minimal for calling an occupation a profession: self-regulation through systematic required training and collegial norms; a base of technical, specialized knowledge (that is usually assessed before entrance to the field—or shortly thereafter); and a code of ethics. Teaching fulfills the last of these three criteria—as teaching is still widely held as a "calling" in our society—where the unwritten code espouses altruistic service. But the other two criteria present problems.

Regarding the first two of the above criteria, the traditions of recruitment, norms of preparation, and conditions of work in schools are such that the claim that teaching is a profession is weakened (Holmes Group, 1986). The Holmes Group report further points out that during the last century, teaching has frequently been a transient career—with young adults teaching school temporarily before assuming the responsibilities of their real careers. Women often chose marriage and full-time housekeeping, while men moved from precollege teaching into higher education or educational management. Teachers (especially in the elementary grades) are overwhelmingly female, while most principals and other administrators are male.

The Task Force on Teaching as a Profession (1986) discusses two additional criteria for characterizing teaching as a profession: autonomy and discretion regarding the organization and content of instruction. Autonomy and discretion, according to the report, are the most attractive aspects of professional work. Schools, however, operate as though consultants, school district experts, textbook authors, trainers, and distant professionals possess more relevant expertise than the teachers in the schools. In the post-Sputnik era of major national curriculum development projects supported by agencies like NSF and the U.S. Office of Education (now the U.S. Department of Education), one frequently heard of "teacherproof" curricula, even though teachers were on the curriculum development project teams. Today, many local and state school officials contend that the teacher's primary role is to deliver the curriculum determined by someone else—and rarely to modify it. This view of teaching hardly allows for autonomy and discretion.

Teachers often complain that the conditions they find in their schools do not allow them to use all the worthwhile skills and knowledge they acquired in their teacher preparation programs. Bureaucratic management of schools, proceeding from the view that teachers lack the talent and motivation to think for themselves, goes against the idea of professional autonomy. Furthermore, the increase in testing as a means of monitoring student progress (and, in turn, teacher and school performance) leads to a narrowing of the curriculum in anticipation of tests. This trend in America is rather ironic, in comparison with trends in other nations. In recent years, England, France, Japan, Kenya, and other countries—which have a long history of testing their students in academic high schools—are broadening their curricula in order to enhance student learning opportunities and teaching flexibility.

The dilemma about autonomy and discretion, and the proper role of the teacher, is further illustrated by the following examples. Case studies (Atkin, 1983) of science and social-science education identified a fourth-grade teacher who was a skillful and talented rock collector. As a consequence, her students could expect to devote a significant portion of their science time to identification and classification. A third-grade teacher was successful at incubating chicken eggs and communicating the intricacies associated with that phenomenon. Her science program consisted mostly of developing, examining, and preserving chick embryos. In addition to these two examples discussed by Stake and Easely (1978), B. Lindsay (personal communication, 1987) reported the example of the social-science teacher who collected dolls during her American and international travels and used them in her classes. The dolls, dressed in clothes of the particular region or country, helped raise many questions about people elsewhere. The school syllabus may not specify rocks in grade 4, chicks in grade 3, or dolls in social-science classes; but for these teachers and their students, science and social-science classes were the most interesting time of the school day. Students and teachers enthusiastically awaited these periods. But there are well-intentioned forces that inhibit this kind of innovation, autonomy, and discretion.

The goal orientation and accountability now being emphasized in schools, reinforced by testing programs, are reasonable and proper. However, they may stifle the idiosyncratic strengths of creative teachers (part of the norms in teaching), for the sake of guarding against teacher weakness. The tension between autonomy and discretion, on the one hand, and accountability, on the other, inhibits efforts to strengthen teaching as a profession.

But another tension also has this effect: the tension between parent and teacher. Parents want the teacher to do what is best for their child. Yet, teachers must take into account the needs of the entire group of students in the classroom and their own teaching styles. For example, one teacher reported a persistent conflict she has with parents regarding her treatment of the periodic table of the elements in her chemistry class. For scientists, the periodic table of the elements is regarded as a great achievement, because its significance rests primarily not on its individual entries, but on the structure and relationships of atoms that the table represents. The teacher reports that despite the great effort she makes to teach the underlying structure and relationships, she and her principal regularly receive complaints that she cannot be a very good chemistry teacher, in that she does not require her students to memorize the periodic table. Parents often think—and perhaps rightly so—that tests require their children to cite the individual entries.

Clearly, there is some question as to whether teaching, as experienced in American classrooms today, is the profession that many think it ought to be. The status of teaching as a profession depends on several factors, such as the relationships among the working conditions of classroom teachers and their need for autonomy and discretion, the need for accountability to parents for the education of their children, and the need for school officials to ensure quality educational standards.

In view of this conflict between the vision of the teacher as a professional and the reality of deprofessionalization of teaching, it is essential that additional steps be taken to strengthen teaching as a profession. The foregoing suggestions for altering university programs for the preparation of teachers so that teachers may increase their specialized knowledge are necessary, but not sufficient.

An interesting mechanism with significant potential for strengthening teachers' identification with a profession can be found in the NSF Science Teacher Institutes. The NSF institutes were begun on an experimental basis in the mid-1950s, were expanded greatly as a result of Sputnik in the late 1950s, became a major force in American science education while reaching their zenith in the 1960s, and were discontinued in the early 1970s. There were two public reasons given for the discontinuation of the NSF institutes: they were no longer needed, and there was a lack of evidence of their effectiveness—where effectiveness was defined in terms of the impact of the institutes on the learning of the pupils of the teachers who attended the institutes (U.S. General Accounting Office, 1984).

The first reason is certainly politically understandable. However, the second reason involves what I call the Winnie the Pooh fallacy that distracts attention from the real significance and impact of the NSF institutes. In one of the Pooh episodes, Pooh wants to capture a heffalump, alleged to be roaming in the forest, and Pooh and Piglet ponder what sort of trap is needed. When the trap is found to be empty the next morning, Pooh and Piglet are unsure whether this means there are no heffalumps in the forest or they just didn't set the right trap. This fallacy of evaluation—if you can't find (or measure) it, then it is not there—is not a sound basis for public-policy formulation.

On the other hand, there is solid evidence of some real impacts of the institutes. For one thing, they promoted the integration of content and pedagogy in science education to a degree unheard of before or since. They were especially strong in the area of pedagogical knowledge of content long before Shulman (1986) identified it. The institutes represented one of the few forces that existed for a focus on subject matter in teaching and on subject-specific pedagogy.

Another important impact of the NSF institutes was strengthening teachers' identification with the professional scientists in the subject fields of their teaching. If teachers left the classroom as a result of enhanced capabilities resulting from institute participation, the evidence is that they stayed in science education, by becoming supervisors or taking on other posts in school leadership or by going into higher education and becoming teachers of teachers. To this day, NSF teacher-institute participants of 20 or more years ago generally exhibit a professional elan not shared by all their colleagues. I believe that the re-establishment of the NSF institutes, especially those promoting full-time, in-depth study during the academic year or summer, would be a significant step in support of the current movement to strengthen teaching as a profession.

ACKNOWLEDGMENTS

It is a pleasure to acknowledge the significant contributions to this paper of Patricia Simmons and Russell Yeany of the University of Georgia College of Education's Department of Science Education.

REFERENCES

Atkin, J. M. 1983. The improvement of science teaching. Daedalus 112: 167-188.

Bandura, A. 1973. Influence of models' reinforcement contingencies on the acquisition of initiative responses. In K. C. Murray and R. Fitzgerald. Interaction analysis modeling and student teacher verbal behaviors. Contemp. Educ. January:174-178.

Buccino, A. 1985. Responding to the condition of science education. Appraisal: Science Books for Young People 18(1):3-15.

Glynn, S., B. Britton, M. Semrud-Clikeman, and K. D. Muth. 1989. Analogical reasoning and problem solving in science textbooks, pp. 383-398. In J. A. Glover, R. R. Ronning, and C. R. Reynolds, Eds. Handbook of Creativity: Assessment, Research, and Theory. New York: Plenum.

Holmes Group. 1986. Tomorrow's Teachers. East Lansing, Mich.: The Holmes Group, Inc.

Kilpatrick, J. 1987. The medical metaphor. Focus Learn. Prob. Math. 9(4):1-13.

Oppenheimer, J. R. 1956. Analogy in science. Amer. Psychol. 11:127-135.

Resnick, L. 1983. Mathematics and science learning: A new conception. Science 220:477-478.

Shulman, L. S. 1986. Those who understand: Knowledge growth in teaching. Educ. Res. 15(2):4-14.

Stake, R., and J. Easley. 1978. Case Studies in Science Education. Washington, D.C.: National Science Foundation.

Task Force on Teaching as a Profession. 1986. A Nation Prepared: Teachers for the 21st Century. New York: Carnegie Forum on Education and the Economy.

Turkle, S. 1984. The Second Self: Computers and the Human Spirit. New York: Simon and Schuster.

U. S. General Accounting Office. 1984. Report on Impact of NSF Teacher Institutes. Washington, D.C.: U. S. General Accounting Office.

24

Biology Teacher Training: Preparing Students for Tomorrow

PATRICIA C. DUNG

We are in the midst of a scientific revolution and critical period in science education. Never before have so many scientific and technological advances been made in so short a time. Only with enough well-trained science teachers are we, as a nation, going to keep up with the demand for top-level scientists, technicians, and educators. But the percentage of high-school students taking 3 years of science across the nation is only approximately 25%, compared with 100% in other industrial nations, such as the U.S.S.R., West Germany, and Japan. Of our nation's students, 84% do not take physics, 65% do not take chemistry, and 23% do not take biology (NSB, 1983). Currently, there is a shortage of science teachers in general and an acute shortage of science teachers with current backgrounds in the subjects they teach (National Journal, 1987).

But in our nation's biology classrooms we are not only training future biologists, technicians, and educators, but also educating the majority, the future citizens of an increasingly technological society and threatened world. The gap between science and technology and the scientific literacy of the average citizen in the United States is ever-widening. As we examine the

Patricia C. Dung is a science adviser in the Office of Instruction, Los Angeles Unified School District. She is also director of Los Angeles Educational Partnership's Target Science Project, a National Science Foundation-funded project, and instructor of the clinical-methods course in secondary-school science at the Graduate School of Education, University of California, Los Angeles. Ms. Dung taught high-school biology for 20 years in the Los Angeles Unified School District.

trends in science proficiency of students from 1969 to 1986, as measured by the National Assessment of Educational Progress, we see that, although there have been some gains over the last 4 years, a majority of high-school students in our country "are poorly equipped for informed citizenship and productive performance in the workplace." The results also indicate that only 7% of high-school students have the knowledge and skills necessary to be successful in college-level science courses (Rothman, 1988). Females and members of minority groups enroll in fewer science courses (Deboer, 1986) and have decidedly lower achievement scores (Thomas, 1986) than males and whites. It is distressing to see that our school systems are producing citizens lacking in the basic understanding of humans' effect on the ecological balance of the Earth, the implications of recombinant-DNA research, and the greenhouse effect, to name a few areas affecting the world as we currently know it. This is a very serious challenge; one that can be met only with motivated and well-trained biology educators in our nation's schools.

Any preservice or in-service training for biology teachers must, by necessity, look at the desired outcome or product, the student, as goals are set and strategies planned. What knowledge and skills must average citizens have to deal with the uncertain future? Certainly they must have an understanding of basic scientific concepts, be able to deal with new information and not be mired in outdated dogma, and be able to relate and apply scientific ideas. Future voters should be able to make informed, intelligent decisions regarding science and society. Future workers, we are told, will have an average of five careers during their lifetimes, and students of the present must be readied for lives and jobs that may not exist today. The future generation will also be the caretakers of the planet.

In order to accomplish our goals, students should experience biological science as a method of solving problems to understand better the place of science in our society. Hands-on, process-based experiences should be a vital and integral part of any high-school biology course (Shymansky, 1984). The best science learning environment is one where students act like real-life, practicing scientists, who engage in a systematic, reciprocal process of theory-building and -testing (Shaw, 1983). Student engagement is necessary for active learning. Biology instruction should emphasize the development of conceptual understanding and critical thinking, rather than the memorization of facts.

Biology teachers must have a model of science instruction that will develop the student skills just described. In the absence of teacher training and analysis, teachers teach as they were taught (Galleger, 1967). Many of our nation's science teachers may not have been presented with good models when they were science students. They may have been students in biology classrooms where they acquired knowledge-based content only, delivered by heavy doses of lectures and textbook work, and did not engage in activities that promoted higher-order thinking skills. If this is true, the preservice and in-service development must provide the missing models, as well as providing appropriate training.

School districts alone cannot provide the growth experience necessary for biology teachers. Schools must be joined by partners from universities, the private sector, and cultural institutions, such as science museums and zoos, in a collaborative effort. Each partner has an important stake in this endeavor. The private sector would like a more informed and skilled work-force in the future; the universities would like better-prepared in-coming students. Appropriate outside resources, when properly coordinated and channeled, can be critical elements in stimulating and facilitating an im-proved instructional environment (NSB, 1983). But the separate partners must do what they do best. Private-sector researchers should not presume to tell teachers how to teach, but can provide important knowledge of technological applications of scientific theory for their colleagues in the classroom.

Universities, being institutions of research, can best provide biology teachers with content updating and research experience and findings. Cur-rent advances in immunology, DNA research, and other areas can be imparted to teachers through precollege teacher institutes and conferences. As indicated earlier, many teachers do not have the specific content knowl-edge of the biology topics they are expected to teach. Also, many biology teachers are lacking in knowledge of the ecology of their immediate envi-ronment, such as the city, and thus lack examples relevant to their students. University professors can provide valuable basic content information to content-deficient teachers through classes and institutes.

Most biology teachers have come through undergraduate science pro-grams without ever having engaged in research activities. The research experience is important in order to convey to students the true nature of scientific research. Programs like those of the Research Corporation and the Industry Initiatives for Science and Mathematics Education-Los An-geles (IISME-LA) match science teachers with university researchers for summer internships that provide experiences that enhance the background of science teachers and produce ideas for classroom laboratory activities. Graduate schools of education can convey current pedagogical research findings that validate excellent teaching practices through seminars for lead teachers and science specialists. Workshops can then be developed that train teachers in these practices. Information on, for example, teacher questioning and wait-time, teacher expectations, sex differences in the sci-ence classroom, and cooperative group learning does not get translated into classroom practice if it is only published and not disseminated through other means.

School district in-service training planners must recognize that teachers learn successful classroom strategies best from other teachers. Every in-service program must provide as models the very behaviors that lead to excellence in biology teaching. Employment of cooperative group learning, hands-on activities, the art of questioning, and other good practices are necessary. Currently, the Los Angeles Unified School District (LAUSD) is implementing a 16-hour advanced-placement biology in-service program with a ratio of one teacher-leader to four less-experienced teacher-participants for small-group discussion and laboratory practice. This low participant-to-leader ratio seems to be promoting greater networking and participation among teachers and the transfer of successful practices from experienced teachers to less-experienced ones.

Another in-service model in the LAUSD, funded by Public Law 98-377, uses lead advanced-placement biology teachers to teach first-year college-level content and process much as they are taught in advanced-placement classes to teachers deficient in particular content areas. Both basic content and good classroom practices are taught.

Workshop in-service programs may provide excellent teaching strategies and models, but an important ingredient is missing: the students. Observing the enthusiasm of students engaged in learning biology provides not only a more complete model, but the stimulus for the observing teachers to try new strategies in their classrooms. Observation of lead or mentor teachers is a most important activity that should continue throughout a teacher's career. Release time should be provided for all teachers to observe other teachers. In the absence of release time, video tapes of exemplary teachers and their classes are helpful (Yeany and Padilla, 1986). Peer or self feedback and analysis without fear of teacher evaluation have also been effective in producing behavioral change (Yeany and Padilla, 1986).

Collaboration between private organizations and school districts can provide the stimulus for educational change. The Los Angeles Educational Partnership (LAEP) was formed as a response to the national educational reform movement. Target Science—funded by the National Science Foundation, the Carnegie Corporation, and private contributors—is a project of LAEP and the LAUSD. Target Science channels science-rich resources to K-12 feeder-school complexes in predominantly minority-group areas. Grants are given to teachers to promote innovative classroom ideas and unique professional development experiences. Museum workshops are held for isolated minority-group parents and students for whom visiting the museum is a unique experience. Dialogues between teachers take place to promote teacher decision-making by assessing the state of science instruction in each feeder complex and forming solutions that lead to an articulated and activity-based K-12 science instructional continuum.

Another Target Science program is IISME-LA, adapted from the successful IISME pilot initiated in San Francisco by the Bay Area industries and the University of California. In Los Angeles, 8-week work experiences at universities and in business and industry are provided for science and mathematics teachers in urban impacted schools. Experience in the private sector provides teachers with real-world examples of practical applications of scientific theory and information about science-related career pathways for students. During this period, teachers actively engage in applications of the science they teach. By so doing, they meet the following objectives: professional development by learning science through the act of engaging in the process itself and stimulation of ideas on how to incorporate real-world experiences to make curriculum more relevant and current. Examples of the research projects that teacher-fellows have participated in are those on the effects of a company product on aortic plaque and on characterizing enamel-making proteins in embryo teeth. Each teacher-fellow is required to develop a curriculum project that can be taken back to the classroom and disseminated to other teachers. Staff time is provided by the company for a mentor for the teacher-fellow. The full cost of employment is funded by the private sector and grants from foundations and organizations, such as National Medical Enterprises.

Target Science also uses private-sector partners as workshop presenters who provide teachers with updated knowledge of technology and science applications, university professors who deliver content, and lead teachers who model hands-on activities and teaching strategies. Topics have included acid rain, urban ecology, immunology, and molecular genetics.

Through TELEventure, another program of Target Science, local industry and university scientists will be "on-line" through an electronic bulletin board with teachers in the Los Angeles area to promote the exchange of information, resources, and activity-based lessons.

All the stakeholders in science education have their own objectives, but it is only through collaboration that we can begin to attack the enormous challenge facing us all.

REFERENCES

Deboer, G. E. 1986. Perceived science ability as a factor in the course selections of men and women in college. J. Res. Sci. Teach. 23:343-350.

Galleger, J. 1967. Teacher variation in concept presentation in BSCS curriculum program. BSCS News. 30:8-19.

National Journal. 1987. The numbers game—statistics on the new school term, a shortage in the classrooms? 19:1996.

NSB (National Science Board Commission on Precollege Education in Mathematics, Science and Technology). 1983. Educating Americans for the 21st Century: A Plan of Action for Improving Mathematics, Science and Technology Education for All American Elementary and Secondary Students. Washington, D.C.: National Science Foundation.

Rothman, R. 1988. Science-achievement levels on test "distressingly low." Educ. Week. 8(4):1,11.

Shaw, T. 1983. The effect of a process oriented science curriculum upon problem solving ability. Sci. Educ. 67:615-623.

Shymansky, J. A. 1984. BSCS programs: Just how effective were they? Am. Biol. Teach. 46:54-57.

Thomas, G. E. 1986. Cultivating the interest of women and minorities in high school mathematics and science. Sci. Educ. 70:31-43.

Yeany, R. H., and M. J. Padilla. 1986. Training science teachers to utilize better teaching strategies: A research synthesis. J. Res. Sci. Teach. 23:85-95.

25

Standards for the Preparation and Certification of Biology Teachers

WILLIAM C. RITZ

Standard: that which is set up and established by authority as a rule for the measurement of quantity, weight, extent, value, or quality [Webster's New Collegiate Dictionary, 1969].

Accreditation and credentialing are procedures through which an agency publicly declares that an institution or individual has or has not met certain standards [Floden, 1988].

INTRODUCTION

Interest in teacher education in the United States seems to emulate a roller coaster ride. Periodically, it becomes a "hot" topic, but then, for long stretches of time, concern about how teachers are prepared appears to stagnate. In the mid-1980s, interest in teacher education emerged from one of its long periods of stagnation when reports of the Holmes Group and the Carnegie Commission raised serious questions about the standards of entry into the profession, what teachers need to know and be able to do, and how we as a nation can ensure an adequate supply of high-quality teachers for our nation's classrooms. Central to much of the debate are

William C. Ritz, director of the Science and Mathematics Education Institute of California State University, Long Beach, serves as president of the Association for the Education of Teachers in Science. He chaired the National Science Teachers Association committee that developed the association's *Standards for the Preparation and Certification of Secondary School Teachers of Science*.

concerns about the standards that are used to decide whether adequate quality exists.

Standards are useless unless applied. Professional standards, such as those dealing with biology teaching, are typically applied to "certification" or "credentialing" processes of one kind or another.

What are the purposes of certification or credentialing? Floden (1988) summarizes them as follows:

> The purposes of credentialing within any profession fall into three major categories: protection of the public, stimulation of improvement, and advancement of the profession. The public is protected if credentialing screens out inadequate schools and incompetent practitioners. Institutions and individuals already meeting minimum standards may be stimulated to further improvement by participation in credentialing procedures. The profession may gain status and other benefits by raising the quality of its membership and by publicly demonstrating responsible intra-professional quality control.

It may be helpful to examine ways in which standards are typically used to attain each of these ends.

USES OF TEACHER PREPARATION AND CERTIFICATION STANDARDS

By and large, the most common and important use of professional standards is to protect the public. With regard to biology teaching, the goals are to ensure that institutions do a proper job of preparing teachers to teach the subject and that those entering the profession will be competent biology teachers. Here, standards are being applied in two very different ways—on the one hand, to assess *institutional* teacher-preparation programs, and on the other, to assess the competence of *individuals* seeking to enter the profession. While state departments of education are usually active in both arenas, we shall see that professional organizations are also assuming important roles in each. Not surprisingly, professional organizations also take a special interest in the other two functions listed by Floden: stimulating further improvement in teacher-education programs and upgrading the status of the science teaching profession by publicly demonstrating responsible "quality control."

THE GREAT DEBATE

Before considering statements of standards adopted by two leading professional groups, it is important to become aware of what some of my colleagues refer to as the debate between the "lumpers" and the "splitters." "Lumpers" are those who espouse what is sometimes referred to as "broad-field certification." In broad-field programs, teachers are prepared or certified to teach several subjects. A clear example here is elementary-school teachers, because they are typically prepared (and certified) to teach many subjects, ranging all the way from mathematics to music. At the high-school level in science, broad-field certification usually means licensing teachers to teach at least two subjects; in extreme cases, it can be a license to teach any and all high-school sciences. I received a New York State certificate some 30 years ago that "certifies" me to teach "all secondary sciences"—biology, chemistry, physics, earth science, and general science. Now, while some may feel qualified to handle all these courses, I certainly do not. If nothing else, my situation clearly illustrates another important point about certification—one can be certified to teach one or more subjects without necessarily also being *qualified* to teach those courses! More about the difference between being "certified" and being "qualified" later.

"Splitters" view preparation and certification as a matter of educating teachers to be subject-matter specialists. Therefore, in splitter states, one is certified to teach a single subject, such as biology. Actually, most splitter states also make provisions for teachers who specialize in one subject to "add on" authorization to teach additional subjects.

The debate between broad- and narrow-field certification continues to this day, and each side has some good points to make. The lumpers point out that most science teachers are expected to teach more than one subject, and they argue the importance of acquiring an interdisciplinary perspective of the sciences. Splitters emphasize the importance of mastery of one's discipline and the belief that one acquires a better perspective of how science works when one specializes in a single discipline. School administrators and boards tend to be lumpers, teacher versatility being especially attractive when you are responsible for making faculty assignments. Subject-matter specialists and many members of the science-teaching profession tend to be splitters, believing in the advantages of specialization. The profession certainly does not speak with one voice with regard to broad-field or more specialized certification.

We see some evidence of the two sides of the debate as we compare two very prestigious statements about the education and certification of biology teachers—those of the National Association of Biology Teachers (NABT) and the National Science Teachers Association (NSTA, 1987). While the standards set forth by NABT (Appendix A) imply a specialization perspective, the organization nonetheless recommends that high-school biology teachers also be prepared to teach "at least one other science" (a splitter statement with a hint of lumping?). The standards of NSTA, on the other hand, clearly favor specialized preparation, even though NSTA separately makes provision for preparation in a second teaching field (Appendix B). Neither organization supports broad-field certification per se. In fact, the NSTA statement reiterates the association's long-standing position that no science teacher ought to be responsible for more than two concurrent course preparations. A close look at the NABT and NSTA statements may prove informative.

I would be remiss if I did not first acknowledge some important predecessors of the current statements: those of the Commission on Undergraduate Education in the Biological Sciences (CUEBS, 1965, 1969) and the American Association for the Advancement of Science (AAAS) Commission on Science Education (1971). Even though we will focus on the more recent statements in this paper, it must be noted that the content of both the CUEBS and AAAS statements remains highly pertinent to this day.

To begin, it should be noted that the NSTA and NABT statements are much more alike than they are different. Both favor strong and broad preparation in biological subject matter (NABT, 24 semester-hours; NSTA, 32), and they specify similar subdisciplines to be studied. Both encourage significant study in related sciences (NABT, 24 semester-hours; NSTA, 16). As already noted, the NABT statement makes explicit its position that all high-school biology teachers should be prepared to teach at least one other science. The NSTA position statement does not speak directly to this notion. (Remember, however, that NSTA does state policies pertaining to the "supplementary authorization" to teach an additional science, and it also acknowledges the need in small schools for science teachers to be able to teach more than two sciences.)

There are some other differences, which may be unintentional. The NSTA statement, for example, makes it explicit that science-teacher preparation ought to address specifically the development of candidates' communication skills, their knowledge of safety in science teaching, and their development of broad research skills—the latter at least to the point of being able to comprehend research results and then communicate them to the public. NABT is less specific about these issues. The considerably more detailed NSTA statement also is more specific about the length and nature of student teaching and other field experiences.

IMPLEMENTATION OF PROFESSIONAL STANDARDS

Standards can be applied as a sort of "metric system" for the processes of accreditation and certification, but in order to serve that function, they must first attain widespread acceptance. The NSTA standards took a giant step forward a few years ago when the National Council for the Accreditation of Teacher Education (NCATE) decided to involve professional organizations, such as NSTA, directly in the accreditation process.

NCATE, the teaching profession's primary mechanism for voluntary regulation, accredits 525 of the nation's 1,276 teacher-education programs. These programs prepare about 80% of our teachers (Padilla, 1988).

Science-teacher preparation programs of colleges seeking NCATE accreditation are now being evaluated by NSTA committees, which use the NSTA standards in their deliberations. This gives the NSTA standards the potential to serve as a powerful tool to strengthen college programs for the preparation of science teachers. It has also caused people to take a much more direct interest in the standards. For example, some who paid very little attention to these standards as they were being developed now find their programs coming up short vis-à-vis those standards. That can be very uncomfortable. No one enjoys being told that his or her program is in some way "below standard." Many have chosen to use the standards as a wedge to bring about changes in their programs. A few have reacted to negative comments about their programs by demanding that the standards be made more compatible with current practices. In developing statements of standards, most organizations believe that, while standards need to be in tune with reality, they should nonetheless work to "stretch" organizations and agencies in directions deemed more desirable. It does little good to set standards so high that no one can attain them. Setting standards at the lowest common denominator is also foolish, wasting, as it does, any opportunity to promote positive change.

The NSTA-NCATE accreditation process is relatively new, but it is moving ahead rapidly. In the fall of 1987, the applications of 45 programs were reviewed, and some 50-60 more are expected to be reviewed in November 1988. While a number of problems have yet to be ironed out, an NSTA-NCATE coordinator has been chosen, and training programs are being conducted for potential reviewers. The exciting part of all this is that a professional organization is now in a position to influence how teachers entering that profession are being prepared.

NSTA'S PROGRAM TO CERTIFY INDIVIDUAL SCIENCE TEACHERS

NSTA is making yet another use of its standards. A few years ago, it began a program that invites individual teachers to submit their educational and professional records to the scrutiny of their science-teaching peers. What does an individual teacher gain from such certification? Two rewards come to mind: recognition from the association (in effect, from one's science-teaching peers) and being able to announce to one's colleagues and school board, "I am an NSTA-certified science teacher!" Those who helped to implement this new program hope that, as an increasingly large nucleus of NSTA-certified teachers develops, states and those hiring teachers will pay more attention to the organization's standards and what they mean—perhaps even adopting those standards as their own.

Recognition as an NSTA-certified teacher may also reduce the possibility of misassignment. The organization plans to support any NSTA-certified teacher threatened by misassignment or replacement by an unqualified teacher by notifying the appropriate school board and local media.

It is NSTA's intent that this program shall be entirely self-supporting, so those seeking individual certification are required to pay a fee (currently $50) for this service. Even though the program is still new, about 150 teachers have been certified (roughly 10% in biology). While this program has yet to have an impact on an impressive proportion of NSTA's membership of almost 50,000, the interest being shown in individual certification seems to indicate that many science teachers see themselves as members of a proud profession and that many of its members want to be recognized for their scholarly and professional attainment.

NSTA is also developing a program of recognition for teachers seeking to claim a still higher level of science-teacher status. While the details are still under development, current thinking involves requiring candidates to have completed at least 5 years of successful full-time science teaching, to submit videotaped science lessons for assessment, to provide acceptable evidence of leadership in science teaching, and, possibly, to attain a stated minimum score on the Graduate Record Examination.

ALTERNATIVE CERTIFICATION PROGRAMS—GOOD NEWS AND BAD NEWS

In recent years, so-called alternative certification programs have received a good deal of publicity. The purposes of such programs are to supply needed numbers of teachers more quickly and to attract promising candidates who might otherwise not enter the profession (Huling-Austin, 1988). Among the states offering such programs are New Jersey, Virginia, California, and Pennsylvania.

What is different about the alternative programs? While the approaches vary, it is possible to identify several characteristics usually found. Most commonly, traditional student-teaching is replaced by entry into a paid, full-time teaching position. The assumption is that carefully selected, mature, confident candidates will be able to move directly into the classroom. That assumption identifies some of the other characteristics one typically finds in alternative programs:

- More stringent selection processes are employed, to identify competent, mature, and confident candidates.
- Participating schools are required to provide on-site support, usually through a mentor teacher.

- A cooperating university provides needed professional education coursework on a reasonable schedule for candidates.
- A university supervisor monitors the candidate's progress through periodic visits and consultation.

In what ways do the candidates for alternative programs differ from those in traditional programs? They tend to be older and to hold a college degree already. Many are migrating to teaching careers from previous careers in business, industry, or the military. Because many candidates are older than their "traditional" counterparts, they tend to bring greater maturity to their entry into the classroom. Often, candidates indicate that they cannot afford to engage in traditional (and unpaid) student-teaching experiences, since they have families to support.

As indicated, alternative programs usually involve accelerated entry into classroom teaching, a feature particularly attractive to many students. Less desirable is the fact that accelerated entry into teaching usually comes at the expense of preparation for the realities of classroom teaching. Candidates often are called on to complete their preparation in methods as they teach. And, while most programs promise on-site support from mentor teachers, such support is often limited.

Since accelerated programs often are available in large urban school districts, candidates frequently end up in "hard-to-staff" schools. Hard-to staff positions receive that designation for a variety of reasons, but it often means serving a particularly difficult population of students. Not unexpectedly, such placements are challenging to even the best and most experienced teachers. The difficulties for one just entering teaching can be overwhelming.

For some, alternative certification programs work very well. They provide early entry into paying positions and tenure tracks. For some candidates—those who are especially mature and self-assured—alternative certification can provide a wonderful match.

Sadly, however, there is evidence that the turnover rate for those entering accelerated programs may be excessively high. Almost half of all teachers entering the profession will leave during the first 7 years. Citing inadequate preparation and on-site support, coupled with placement in difficult teaching situations, Huling-Austin (1988) speculates that the dropout rate for "accelerated" teachers is even higher.

The programs that work well are often those providing the strongest assistance and support to candidates. Huling-Austin makes an excellent case for the types of support needed to increase the retention of teachers. She cites the need for careful placement, appropriate on-site support, and careful attention to meeting the specific needs of the individuals involved in the program as key factors determining success. My own experiences with the accelerated program at California State University, Long Beach are in tune with the points made by Huling-Austin. With early placement of carefully selected candidates in settings that nurture their development as science teachers, alternative certification can work very well. To the extent that these conditions are not met, we risk wasting promising candidates who might under better circumstances develop into excellent science teachers.

QUO VADIS? ISSUES TO BE ADDRESSED AND RESOLVED

Despite the progress that is being made toward ensuring that every high-school science student will be taught by a fully qualified teacher, a number of serious problems remain. Among them are the following:

- In the absence of a firm research base, the impossibility of knowing whether *anyone's* standards, if met, will ensure high-quality science teaching.
- Sometimes weak state standards for science-teaching credentials.
- The misassignment of teachers and the use of "emergency" credentials.
- The use of standardized tests as the sole determinant of subject-matter competence.
- The continuing need to re-evaluate and rewrite standards.
- The limitations of standards to effect real change.

A serious problem with all standards is the lack of data that might tell us whether one set of standards is significantly better than any other. The analogy of medical and legal credentialing practices is often raised, but Floden (1988) points out that even there, very little evidence is available to support the credentialing of these professions. Be that as it may, most of us hold an intrinsic belief that there exists some minimum of preparation below which satisfactory performance as a science teacher is highly unlikely. In the case of biology teaching, is that 20 semester-hours of preparation? 30? 40? Whatever that number is, what sort of distribution of study in biology should exist? Is it imperative that every biology teacher assimilate the equivalent of two courses in botany? What about embryology? Evolution? We simply do not now have research-based answers to these questions, even though statements of standards exist and those standards are being used to make important decisions about who shall and who shall not teach biology. Obviously, then, there exists a great need for data on which a rational set of standards can be based.

Whether one is philosophically a so-called lumper or a splitter with regard to the preparation and certification of teachers, all professional scientists and science educators should take an active interest in what is happening in their own states. Are the standards adequate? Do they fall significantly short? While most of us would agree that the typical high-school biology teacher does not need to have a Ph.D., we would also tend to agree that there is some minimum amount of education in the life sciences that ought to have been attained. If your state's standards are significantly less demanding than those of the NABT or NSTA, you and your colleagues ought to see what might be done about that situation.

Increasingly, states are turning to standardized testing as a major tool to determine who does and who does not get a teaching credential. Some have greater faith in the efficacy of standardized testing than do others, including me. Should the results of a standardized test determine who may and who may not teach high-school biology? Is course- and credit-counting more reliable than test results? Most of us have our own beliefs with regard to all this, but we have alarmingly few data to support our views, whatever they may be.

As to misassignment of teachers, school districts can and frequently do assign teachers to classes that they are both uncertified and unqualified to teach. This practice tends to occur most frequently during times of teacher shortages. Weiss reported in 1987 that some 11% of high-school biology classes are being taught by teachers who did not major in biology. In a reanalysis of the data, however, she found that 80% of the nation's biology teachers have actually completed at least eight courses in their discipline (Weiss, 1988a). Despite this seemingly good news, however, only about one-third of biology teachers meet all the NSTA standards referred to earlier. It appears from Weiss's more recent study (1988b) that misassignment in science most typically places science teachers in science classes for which they are but partially prepared. Weiss estimates that some 6% of high-school science classes are taught by teachers with very little science background. She notes that misassignment is less likely to occur in biology than in chemistry or physics classrooms.

It is easy to criticize teacher misassignment, but problems at the local level can make avoiding the practice very difficult at times. If you were the principal of a high school, and you were short one biology teacher, would you or wouldn't you assign another teacher, unfortunately not certified, who was available? Your rationale might be that that teacher did actually take a few biology courses as an undergraduate. And isn't a course called "Exercise Physiology" really a life-science course, even though it was not offered through the biology department? And if we don't have this person teach one or more science courses, we might even have to let her or him go! The forces that operate in favor of misassignment are indeed strong and difficult to counter. At the very least, however, we ought to insist that school districts that misassign a teacher into a biology classroom be obliged to help that teacher to become properly qualified for that assignment.

One of the difficulties we encounter once standards have been written is that there is a tendency to think of them as being "finished." There are some very understandable reasons for this. It is not much fun to sit down and write standards. These tasks are usually done by a committee, and that implies a need to achieve consensus. Even worse, they are usually created under the auspices of an organization, and achieving consensus under these conditions is even more difficult.

And so, once standards have been written, have passed the scrutiny of an organization's members, and have finally appeared in print, few persons have the courage or determination to continue the developmental process. And yet standards, once written, must continue to be evaluated and, as necessary, rewritten. To the extent that this does not happen, we increase the odds that the standards will fail in their intended purpose.

It is always vital to keep in mind the things that standards cannot do. Establishment of standards today does not mean that the problems addressed by those standards will go away tomorrow. When change in organizations and bureaucracies does occur, it typically can occur but slowly. Standards are tools that can be put to good use as we work together to improve the quality of high-school biology teachers, but it takes much work, over a great deal of time, for those changes to occur.

REFERENCES

AAAS (American Association for the Advancement of Science). 1971. Guidelines and Standards for the Education of Secondary School Teachers of Science and Mathematics. Miscellaneous Publication 71-9. Washington, D.C.: AAAS.

CUEBS (Commission on Undergraduate Education in the Biological Sciences). 1965. Preparing the Modern Biology Teacher. Position Paper of the Panel on Preparation of Biology Teachers. Reprinted in BioScience 15:769-772.

CUEBS (Commission on Undergraduate Education in the Biological Sciences). 1969. The Pre-service Preparation of Secondary School Biology Teachers. A. Lee, Ed. CUEBS Publication 25. Washington, D.C.: The George Washington University.

Floden, E. 1988. Analogy and credentialing, pp. 13-19. In Action in Teacher Education, Spring-Summer 1979. In Action in Teacher Education: Tenth-Year Anniversary Issue, Commemorative Edition. Reston, Va.: Association of Teacher Educators.

Huling-Austin, L. 1988. Factors to consider in alternative certification programs: What can be learned from teacher induction research?, pp. 169-176. In Action in Teacher Education: Tenth-Year Anniversary Issue, Commemorative Edition. Reston, Va.: Association of Teacher Educators.

National Association of Biology Teachers. Undated. NABT Biology Teaching Standards. Arlington, Va.: National Association of Biology Teachers.

National Science Teachers Association. 1987. Standards for the Preparation and Certification of Secondary School Teachers of Science. Washington, D.C.: National Science Teachers Association.

Padilla, M. J. 1988. Using the NSTA Teacher Education Standards: Preparing NCATE Folios Applying for NSTA Certification. Washington, D.C.: National Science Teachers Association.

Webster's Seventh New Collegiate Dictionary. 1969. Springfield, Mass.: G. & C. Merriam Co., Publishers.

Weiss, I. R. 1988a. Course Background Preparation of Science Teachers. Paper prepared for the AAAS Forum on School Science, 1987. Data quoted in The Present Opportunity in Education (p. 6), a position paper of the Triangle Coalition for Science and Technology Education, September 1988.

Weiss, I. R. 1988b. Course background preparation of science teachers in the United States: Some policy implications, pp. 97-118. In A. B. Champagne, Ed. Science Teaching: Making the System Work. Washington, D.C.: American Association for the Advancement of Science.

APPENDIX A

NABT Biology Teaching Standards

PROGRAM SCOPE: A biology education program should prepare teachers for both the junior high/middle school and senior high school levels of instruction and should be designed to educate college and university students to teach any secondary biology or other life science courses. The suggested program should include a minimum of 24 semester hours in the biological sciences, including course work to ensure the proficiencies stated below; plus a minimum of 24 semester hours in chemistry and introductory physics; and proficiency in mathematics through college algebra. A minimum of 12 semester hours in science should be upper division hours. All secondary biology teachers should be prepared to teach in at least one other science area. In addition, biology teachers must continue to improve their skills and knowledge in the ever changing world of life sciences.

Standards For Preparation of the Biology Teacher

Candidates completing a biology teacher certificate program must include the curricular goals listed below and be able to demonstrate specific skills and knowledge.

1. Knowledge of the fundamentals of Biology.
   - Demonstrate a knowledge of basic concepts and of laboratory techniques concerned with the study of: systematics; development; evolution; genetics; ethical implications of technology (recombinant DNA, organ transplant, in vitro fertilization); ecology; behavior; cell biology; bio-energetics; homeostatic mechanisms; and all the life processes in animals, plants and microbes.
2. Knowledge of the interrelationships of living organisms with their biotic and physical environments, including field experiences and the study of ecology or environmental biology.
   - Demonstrate in writing a knowledge of the basic concepts of ecological population factors; ecosystems; energy flow; nutrient cycles and the sociobiological aspects of ecology.
   - Demonstrate an ability to conduct and direct meaningful field trips and investigations concerned with obtaining information on concepts of ecological populations; ecosystems; energy flow; nutrient cycles and the sociobiological aspects of ecology.
3. Knowledge of chemistry, mathematics, and physical science or physics, and computer science —
   - Demonstrate: a basic knowledge of the concepts and a command of the laboratory techniques equivalent to those included in general college chemistry; the concepts equivalent to those included in lower division, undergraduate physical science course, or a college physics I course; a command, or working ability of mathematics equivalent to that in college algebra; and an ability to utilize computers in teaching and in record storage.
4. A methods course for biology teaching designed to organize, plan, present, and evaluate the learning of biology subject matter content.
   - Demonstrate a functional knowledge of the science inquiry processes and be able to distinguish between assumptions, hypotheses, theories, data, controls, independent and dependent variables, and generalizations.
   - Define and describe a philosophy of present-day science teaching.
   - Demonstrate a command of the mechanics of everyday teaching, including laboratory and field experiences.
   - Select, purchase, operate, and maintain equipment and supplies used in teaching biology.
   - Use current biology curricular materials in the classroom.
   - Demonstrate an ability to develop curricula that motivate students as well as consider individual differences.
   - Demonstrate an ability to construct and administer student evaluation instruments for subject matter concepts, principles, and techniques.
   - Demonstrate a commitment and dedication to education of early adolescents and continual self-improvement.
   - Foster enthusiasm about biology in students of diverse backgrounds.
   - Demonstrate interest in professional growth by actively participating in local, regional, or national biology association programs.

Standards for Professional Growth of the Biology Teacher

Teachers who wish to maintain their skills and knowledge gained in undergraduate work must include the following goals in their profession.

1. Maintain standard of excellence and broaden knowledge of life sciences.
   - Demonstrate professionalism by participating in a biological science teacher education program which will lead to a higher degree.
   - Participate in biology inservice programs and/or summer institutes to learn new teaching methods and laboratory techniques.
   - Participate in local, regional, and national biology conferences to keep abreast of new trends and discoveries.
   - Demonstrate commitment to learning by reading professional journals.
2. Establish close relationship with scientific community, businesses, and industries.
   - Demonstrate interest in scientific community by participating in local and national biology organizations.
   - Develop communication with local businesses, nonprofit organizations and private institutions.
   - Demonstrate leadership by taking active role in maintaining scientific integrity in the community and by sharing biology teaching ideas with colleagues.

APPENDIX B

Standards for the Preparation And Certification of Secondary School Teachers of Science

I. Science Content Preparation

The program for preparing secondary school teachers of science should require specialization in one of the sciences (i.e. preparation equivalent to the bachelor's level) as well as supporting course work in other areas of science. The programs should require a minimum of 50 semester hours of course work in one or more of the sciences and additional course work in related content areas such as mathematics, statistics, and computer applications to science teaching. The programs and courses should be designed to develop a breadth of scientific literacy that will provide the preservice teacher with

- positive attitudes toward science and an accompanying motivation to be a lifelong learner in science;
- competency in using the processes of science common to all scientific disciplines, including the skills of investigating scientific phenomena, interpreting the findings, and communicating results;
- competency in a broad range of research, laboratory and field skills;
- knowledge of scientific concepts and principles and their applications in technology and society;
- an understanding of the relationship between science, technology, society and human values; and
- decision-making and value-analysis skills for use in solving science-related problems in society.

Overall, the programs should be designed for the unique needs of secondary school science teachers.

II. Science Teaching Preparation

Science Teaching Methods and Curricula

The program should prepare preservice teachers in the methods and curricula of science. Method courses should model desired teaching behavior in the secondary classroom. These experiences should develop a wide variety of skills, including those which help preservice science teachers to

- teach science processes, attitudes, and content to learners with a wide range of abilities and socio-economic and ethnic backgrounds;
- become knowledgeable of a broad range of secondary school science curricula, instructional strategies and materials, as well as how to select those best suited for a given teaching and learning situation;
- become proficient in constructing and using a broad variety of science evaluation tools and strategies; and
- become knowledgeable about the learning process, how people learn science, and how related research findings can be applied for more effective science teaching.

The program should include *at least* one separate course (3-5 semester hours), and preferably more, in science teaching methods and curricula.

Communication Skills and Classroom Management Techniques

The program should prepare preservice teachers to speak and write effectively and demonstrate effective

NATIONAL SCIENCE TEACHERS ASSOCIATION,
1742 CONNECTICUT AVENUE, N.W., WASHINGTON, D.C. 20009

use of classroom management techniques when teaching laboratory activities, leading class discussions, conducting field trips, and carrying out daily classroom instruction in science.

Preparation in Research Skills

The program should prepare preservice teachers to conduct or apply, understand, and interpret science education research and to communicate information about such research to others (e.g., students, teachers and parents).

Safety in Science Teaching

The program should require experiences that develop the ability to identify, establish, and maintain the highest level of safety in classrooms, stockrooms, laboratories, and other areas used for science instruction.

Other Educational Experiences

Courses in other educational areas, including general curricula and methods, educational psychology, foundations and the special needs of exceptional students, should be a part of the program in order to complement the science education components described above.

III. Classroom Experience

Field Experience

Field experiences in secondary school science classrooms are essential for the thorough preparation of preservice teachers of science. The field experience of preservice teachers should begin early with an emphasis on observation, participation, and tutoring, and should progress from small to large group instruction.

The Student Teaching Experience

The student teaching experience should be full-time for a minimum of 10 weeks. The program should require student teaching at more than one educational level (such as junior high school experience combined with that of working in the high school) or in more than one area of science (i.e., biology and chemistry) if certification is sought in more than one area. The program should give prospective teachers experience with a full range of in-school activities and responsibilities.

Day-to-day supervision of the student teacher should be done by an experienced, master science teacher(s). University supervision should be provided by a person having significant secondary school science teaching experience. Responsibility for working with student teachers should be given only to highly qualified, committed individuals, and close and continuing cooperation between school and university is imperative.

IV. Supportive Preparation in Mathematics, Statistics, and Computer Use

The program should require competencies in

- mathematics as specified for each discipline;
- scientific and educational use and interpretation of statistics; and
- computer applications to science teaching, emphasizing computer tools such as: (a) computation, (b) interfacing with lab experiences and equipment, (c) processing information, (d) testing and creating models, and (e) describing processes, procedures, and algorithms.

NATIONAL SCIENCE TEACHERS ASSOCIATION,
1742 CONNECTICUT AVENUE, N.W., WASHINGTON, D.C. 20009

Standards for Each Secondary Discipline

Biology

I. The program in biology should require broad study and experiences with living organisms. These studies should include use of experimental methods of inquiry in the laboratory and field and applications of biology to technology and society.

II. The program would require a minimum of 32 semester hours of study in biology to include at least the equivalent of three semester hours in each of the following: zoology, botany, physiology, genetics, ecology, microbiology, cell biology/biochemistry, and evolution; interrelationships among these areas should be emphasized throughout.

III. The program should require a minimum of 16 semester hours of study in chemistry, physics, and earth science emphasizing their relationships to biology.

IV. The program should require the study of mathematics, at least to the pre-calculus level.

V. The program of study for preservice biology teachers should provide opportunities for studying the interaction of biology and technology and the ethical and human implications of such developments as genetic screening and engineering, cloning, and human organ transplantation.

VI. The program should require experiences in designing, developing, and evaluating laboratory and field instructional activities, and in using special skills and techniques with equipment, facilities, and specimens that support and enhance curricula and instruction in biology.

NATIONAL SCIENCE TEACHERS ASSOCIATION,
1742 CONNECTICUT AVENUE, N.W., WASHINGTON, D.C. 20009

26

Current Issues in Biology Education for Teachers

EXYIE C. RYDER

Biology education currently faces several critical issues, particularly in the area of teacher preparation. Like other programs in education, biology education will most probably be affected by recent calls for reform in the teaching profession by the Holmes Group, the Carnegie Task Force on Teaching as a Profession, the National Council for Accreditation of Teacher Education (NCATE), the National Science Board, and others.

It is the purpose of this paper to discuss two current major issues in biology education for teachers. The first is the biology-content component within the undergraduate teacher-education curriculum, and the second is the Holmes Group report and its potential effects on the recruitment of teachers from minority groups.

THE BIOLOGY-CONTENT COMPONENT OF THE TEACHER-EDUCATION CURRICULUM

The preparation of quality biology teachers must include a solid foundation in biology content. Prescribed courses of study should provide breadth of the basic concepts and principles on which the discipline of biology is built, but must also concentrate on the depth of knowledge

Exyie C. Ryder is professor of biological sciences at Southern University in Louisiana. Her primary interest is in biological science education. She holds a Ph.D. from the University of Michigan.

available in the subject-matter field. For many years, the academic subject-matter component of teacher-education programs has come under scrutiny; however, recent outcries for improving the quality of teaching and the teaching profession have raised new concerns over this issue. These reports call for teachers to demonstrate competence in academic subjects and for institutions of higher learning to "make the education of teachers more intellectually solid" (Holmes Group, 1986).

It has been stated by Cadenhead that teaching as an intellectual activity should include knowledge, linking content and methodology, and sentences (Cadenhead, 1985). This statement suggests that a quality teacher-education program in biology should include courses in biology content, scientific methods, general education, and liberal studies. There is no consensus on the proportion of the curriculum or the number of credit hours that should account for each of these four areas. Consequently, there are wide variations in courses required in different curricula. A primary reason for the inconsistencies is the fact that the teacher-education curricula are usually developed around each state's unique certification requirements. Since most states' requirements for certification lean heavily toward the professional-education component, rather than the subject-matter component, the result is that teacher-education programs tend to be long on professional education, including pedagogy, and short on subject-matter content. This condition has prompted critics of the teaching profession and those involved in the current reform movement to recommend that prospective teachers earn a baccalaureate degree in their subject-matter area before being allowed to enter a professional teacher-education program (Holmes Group, 1986; Carnegie Forum on Education and the Economy, 1986).

It is important that a teacher thoroughly understand a subject in order to teach it effectively. This idea is aptly expressed by Murray (1986), who states:

> The teacher's role is to find and present the most powerful and generative ideas of a discipline in a way that preserves its integrity and leads to student understanding. This implies that a teacher comprehends the structure of the discipline, its key points and their origins, and the criteria by which one distinguishes the important from the trivial. This kind of understanding, slighted in traditional programs, is of fundamental importance to the teacher—and must have a central place in the teacher's education.

He further states that "the traditional major often does not confer a level of understanding that empowers the teacher (or even the typical college graduate) to understand" (Murray, 1986).

Breadth and Depth of Content

Within the biology departments of colleges and universities, a prospective teacher of biology should acquire a thorough, up-to-date grasp of the subject matter. This can be achieved by following a curriculum that provides state-of-the-art content, state-of-the-art laboratory skills, and state-of-the-art biological research techniques that are acquired only in a research laboratory setting. The combined experiences offered the students in the lecture, the laboratory, and the research environment will enable preservice teachers to gain confidence in their ability to "do and perform the subject matter, and not just talk about it" (Murray, 1986).

Before any attempts to reform the subject-matter component of the biology teacher-education program are made, several questions regarding the nature of such changes must be asked. For example: Is it really necessary, as some advocates feel, to earn a bachelor's degree in biology before being admitted to a biology teaching program? Who should decide what content and experiences within the biology department will be most meaningful for students majoring in biology education? Should there be increased breadth or increased depth in the coverage of the content? Who should determine whether a teacher has acquired adequate mastery of the discipline?

Ideally, the answers to these questions should be derived from the collaborative efforts of the faculty in education and the faculty in the biological sciences, for the responsibility for preparing high-school biology teachers should be shared by the two groups. For too long, cooperation and collaboration between the subject-matter faculty and the education faculty, with respect to teacher training, have been minimal. What is needed now is a biology faculty that is sensitive to the problems and needs of high-school biology teachers, particularly since the high-school biology teachers prepare the next generation of college biology majors.

The biology department faculty can contribute its subject-matter expertise, its knowledge of the structure of the discipline, and its knowledge and understanding of contemporary topics of research and investigation. The biology faculty can also assist in the identification of a core of courses that will provide the necessary breadth and depth of content in the biological sciences and can recommend a sequence of advanced-level courses that will expand prospective teachers' knowledge base and simultaneously offer enough depth in biology to give the students a high degree of proficiency in biology content and in laboratory skills and techniques. The biology faculty should encourage biology-education majors to develop research skills as an integral part of their training. This could easily be accomplished if the faculty engaged in research projects would use education majors as research assistants in the same way that they use noneducation science majors. In summary, the biology departments must be more responsive to the needs of teacher-education majors, and they must develop greater respect for the role that high-school biology teachers play in preparing students to pursue careers in biology, medicine, allied health, and related fields.

The "Content" Issue and the Curriculum

Since the present 4-year teacher-preparation curriculum is already crowded, how can the content component of the program be strengthened? Proponents of curriculum reforms in the teaching profession favor moving teacher education to postbaccalaureate status and leaving the 4-year undergraduate program for content specialization, general education, and liberal studies. On the other hand, there are those who recommend reorganizing the present 4-year baccalaureate teacher-education program, with a view to eliminating the redundancy in professional-education courses, thereby leaving space in the curriculum to augment the subject-matter area. Alan Tom, a proponent of redesigning the 4-year curriculum, argues against establishing a postbaccalaureate professional school of education, for he feels that doing so "tends to artificially separate the academic and the professional aspects of teaching" (Tom, 1986).

In the revised NCATE-approved curriculum guidelines for biology teacher-education programs, prepared by the National Science Teachers Association (NSTA), it is recommended that high-school biology teachers complete a minimum of 32 semester-hours in biology, with the necessary support courses in other sciences, mathematics, and computer science (NSTA, 1987). The guidelines specify the biology courses that should be included and point out that the approved curriculum gives the biology teacher-education major the content "preparation equivalent to the bachelor's level" (NSTA, 1987). The revised NCATE-approved guidelines for high-school teachers are as follows:

I. The program in biology should require broad study and experiences with living organisms. These studies should include use of experimental methods of inquiry in the laboratory and field and applications of biology to technology and society.

II. The program would require a minimum of 32 semester hours of study in biology to include at least the equivalent of three semester hours in each of the following: zoology, botany, physiology, genetics, ecology, microbiology, cell biology/biochemistry, and evolution; interrelationships among these areas should be emphasized throughout.

III. The program should require a minimum of 16 semester hours of study in chemistry, physics, and earth science emphasizing their relationships to biology.

IV. The program should require the study of mathematics, at least to the precalculus level.

V. The program of study should provide opportunities for studying the interaction of biology and technology and the ethical and human implications of such developments as genetic screening and engineering, cloning, and human organ transplantation.

VI. The program should require experiences in designing, developing, and evaluating laboratory and field instructional activities, and in using special skills and techniques that support and enhance curricula and instruction in biology.

In concluding the discussion of the first major issue, I strongly suggest that each institution that prepares biology teachers consider establishing a biology teacher-education council consisting of faculty from precessional education, science education, and the biological sciences. The council would be responsible for periodically reviewing the biology teacher-education curriculum to ensure a solid foundation in biology content.

POTENTIAL EFFECTS OF THE HOLMES GROUP REPORT ON THE RECRUITMENT OF TEACHERS FROM MINORITY GROUPS

Overview of the Holmes Group Proposals

The Holmes Group report, *Tomorrow's Teachers* (Holmes Group, 1986), is one of several recent reports that propose major reforms in teaching and in teacher preparation. Among the recommendations in the report for improving the teaching profession are the following:

- To abolish the undergraduate-degree program in education and institute a 5- or 6-year program of study as a prerequisite for certification, licensing, and entry into the profession.
- To establish a three-tier system within the profession that would identify and recognize differences in levels of knowledge, skills, and commitment among teachers.
- To create standards of certification to monitor entry into the profession.

There is widespread feeling among minority groups, and nonminority groups as well, that the reform agenda, if implemented, would decrease the number of minority-group members entering teaching and teacher-education programs.

Implications of the Recommendations

The recommendations of the Holmes Group come at a time when the teaching profession is becoming less and less attractive. For years, women and minority-group members staffed the nation's classrooms when opportunities for higher-paying, more attractive positions were unavailable to them. Since many of the barriers to other occupations have been removed, minority-group members and women are opting for careers other than teaching.

The minority-group teaching force in the United States is dwindling—ironically, at a time when the number of minority-group students in the schools is increasing significantly. By the year 1990, members of minority groups could constitute 30% of the American school population. According to Shirley Malcom, of the Office of Opportunity in Science of the American Association for the Advancement of Science, blacks "are projected to account for only 5 percent of the teaching force by 1990" (Jacobson, 1986). Hispanics and members of other minority groups are expected to account for approximately 3% of the teaching force (Haberman, 1988). It is estimated that the nation will need more than 200,000 new teachers by the year 2000. Many of these teachers will be needed in the areas of science and mathematics, where the shortage is predicted to be very acute.

The Holmes Group report recommends extending the period of study for persons entering teacher-education programs. The impact of this recommendation on minority-group teacher recruitment would be devastating, for lengthening the period of schooling would add substantially to the cost of a college education and could result in severe financial setbacks for most minority-group students and their families. Clearly, the implementation of a 5- or 6-year curriculum model would be a deterrent to many minority-group students who might be contemplating teaching, and their reluctance to commit themselves to a career that offers little financial reward is understandable. In short, the prolonged study period would severely hamper the recruitment of minority-group members into the teaching profession. On the other hand, the extended programs could be made attractive to minority groups if assistance in the form of stipends, grants, fellowships, scholarships, and loan-forgiveness programs were made available.

With teacher shortages at a crucial level and expected to rise continuously, it could be argued that the diminishing pool of qualified teachers could be offset if steps were taken to identify a larger body of prospective teachers and provide the necessary academic and financial support for their education and training. To the contrary, it is felt that reforms outlined by the Holmes Group and other commissions will create a very narrow pool of prospective teachers who can afford to elevate themselves (through additional education and training) to the top of the profession.

The Holmes Group's recommendation regarding the establishment of a three-tier system within the teaching profession is also expected to have a negative impact on minority-group teacher recruitment. Minority groups view with skepticism the career ladder with its built-in "hurdles" for advancement. In particular, the vagueness of the phrasing in the report is a matter of concern to many. For example, Beverly Gordon, in referring to the Holmes Group recommendation "to recognize differences in teachers' knowledge, skills, and commitment in their education, certification, and work," points out that minority groups must, in fact, be sure that the so-called differences do not "translate into deficiencies" (Gordon, 1988). Another critic of the career-development proposal calls attention to the fact that "race is a critical variable in any career development scheme" (Oliver, 1988) and notes that the increased emphasis on examination, the extended study period required, and the higher standards for certification all tend to discourage minority-group members from entering the teaching profession.

The Holmes Group also proposes the creation of standards of entry into the teaching profession. While higher standards are desirable and necessary, there is apprehension among minority groups with respect to the standards that are to be created and how they will be applied. Over the last few years, the nation has witnessed the effects of competence testing on minority groups. The result has been the elimination of large numbers of members of those groups from teaching and from entering the teaching profession. An alarming example of the impact of extensive testing is that which has occurred in Florida and 18 other states where testing is apparently the primary reason for the reduction in the minority-group teaching force since the early 1980s (Smith, 1988).

It is estimated that if the Holmes Group proposal to create standards of entry into the teaching profession is adopted and implemented on a national level, 50-85% of minority-group members will be eliminated from teaching. These eliminations will occur through testing, assessment of on-the-job performance, and other forms of evaluations, if the evaluation instruments are developed and validated using the same procedures that have been used previously and if minority groups are not involved in the test development and validation processes (Smith, 1988). Clearly, this trend must be reversed, as ways are sought to attract and retain minority-group teachers.

Minority-Group Teacher Recruitment: The Need and Some Proposed Solutions

The most important reason why minority-group teachers must be recruited is that they are needed in the classrooms as role models for minority-group students. As cited previously, minority-group enrollment in the schools is rapidly increasing, while the supply of minority-group teachers is steadily declining. This situation has resulted in fewer role models for minority-group students, who now account for more than 50% of the enrollment in most of the largest school districts in the country and who are expected to account for more than 38% of the school population in the United States by the year 2000. The presence of minority-group role models is important, because they provide a psychological support system in the schools for minority-group youth and because they are important in the development of those students' self-esteem.

Minority-group teachers are needed in the schools for yet another reason: they contribute to the diversity of the teaching profession. Diversity is a factor that is valued in America's "melting pot," because it allows people of various backgrounds and cultures to interact and learn to appreciate and respect each other and their differences.

To offset the potential effects of the Holmes Group recommendations on the recruitment of minority-group teachers, I propose several solutions:

- Provide incentives to attract minority-group students into the teaching profession. Monetary incentives—such as scholarships, stipends, assistantships, grants, fellowships, and loan-forgiveness programs—would be most desirable.
- Identify a pool of prospective, talented, minority-group teaching candidates and involve them in an academic intervention program that will enable them to enhance their academic skills and improve their test-taking skills.
- Involve more minority-group institutions in the planning for the reforms in teacher education.
- Involve minority groups in the construction and validation of teacher-evaluation instruments.
- Raise the salaries of teachers.

The need for minority-group science teachers is as important as the need for minority-group teachers in general, for minority-group science teachers serve as scientist role models for minority-group students. Therefore, efforts must be made to recruit minority-group members into science teaching.

The recruitment of minority-group members into science teaching must begin with attracting youth to the sciences and then attract them into science teaching. This should be initiated as early as middle school and junior high school, when minority-group youngsters should be encouraged and challenged to enroll in science and mathematics courses beyond those which are required for everyone. An early start will enable the students to develop interest in the sciences and at the same time obtain the prerequisites necessary for success in higher-level science courses.

I strongly suggest that the Holmes Group report and its potential effects on minority-group teachers in general and minority-group science teachers in particular be critically examined.

REFERENCES

Cadenhead, K. 1985. Is substantive change in teacher education possible? J. Teach. Educ. 36(4):17-21.

Carnegie Forum on Education and the Economy. 1986. A Nation Prepared: Teachers for the 21st Century. The Report of the Task Force on Teaching as a Profession. New York: Carnegie Corp.

Gordon, B. 1988. Implicit assumptions of the Holmes and Carnegie reports: A view from an African-American perspective. J. Neg. Educ. 57:141-158.

Haberman, M. 1988. Proposals for recruiting minority teachers: Promising practices and attractive detours. J. Teach. Educ. 39(4):38-41.

Holmes Group. 1986. Tomorrow's Teachers: A Report of the Holmes Group. East Lansing, Mich.: Holmes Group, Inc.

Jacobson, R. 1986. Carnegie school-reform goals hailed: Achieving them called tall order. Chron. High. Educ. 32:1-23.

Murray, F. B. 1986. Teacher education: Words of caution about popular reforms. Change 18:16-25.

NSTA (National Science Teachers Association). 1987. NCATE-Approved Curriculum Guidelines for Biology Teacher Education Programs. Washington, D.C.: NSTA.

Oliver, B. 1988. Structuring the teaching force: Will minority teachers suffer? J. Neg. Educ. 57:159-165.

Smith, G. P. 1988. Tomorrow's white teachers: A response to the Holmes Group. J. Neg. Educ. 57:178-194.

Tom, A. 1986. The Case for Maintaining Teacher Education at the Undergraduate Level. Paper prepared for the Coalition of Teacher Education Programs. St. Louis, Mo.: Washington University.

PART VI

Accomplishing Curricular Changes—Institutional Barriers

27

Educational Reform? Are We Serious? No, but We Had Better Be.

JOHN A. MOORE

In a recent editorial, Koshland (1988) had this to say:

> The nation is intoxicated with huffing, puffing, and crocodile tears as a substitute for policy in the war on drugs.

With minor modifications, that *cri de coeur* characterizes our problem:

> The nation is intoxicated with huffing, puffing, and crocodile tears as a substitute for policy on educational reform.

The general solutions to our problems are obvious and have been for years. We know what needs to be done, but so far there is a pervasive unwillingness to make the necessary changes in the educational establishment to achieve the ends said to be desired. The welfare of the nation requires students with a willingness to learn; teachers fully capable of stimulating and supporting that learning; excellent textbooks, educational equipment, and facilities; political leaders with courage, vision, and ability; and, above all, a society willing to make the sacrifices that will produce the educational system the nation deserves.

But every segment of the educational establishment is inadequate to some degree, and that means that every segment is to some degree a barrier

---

John A. Moore is professor of biology, emeritus, University of California, Riverside. He led the team that developed the yellow version of the Biological Sciences Curriculum Study (BSCS) biology text in the 1960s; is a director of the Science as a Way of Knowing project of the American Society of Zoologists; and is a member of the National Academy of Sciences.

to educational reform. It also means that acceptable and sustainable reform will require a fundamental change in us—for we are an integral part of the educational system.

We are all well-meaning, of course. We gather regularly in meetings like this and issue ukases on what should and must be done. The fact that we repeat again and again what a series of similar committees have been saying for years reassures us that we are "on the right track."

Although there has been a tireless and tiresome listing of what should be done, little effort has gone into doing it. Our task should be to bridge the gap between rhetoric and response. And we have a chance—we are asked to advise the Howard Hughes Medical Institute, which has the resources to take effective action.

This long history of huffing and puffing raises an interesting question: Are we really serious and willing to work for educational reform, or are these many meetings just another example of one of academe's favorite devices—study the problem until it goes away or until the next problem takes its place?

I am not at all sure that the academic community has the stomach to undertake what must be done. Really effective reform would be so difficult and so pervasive that many will elect to settle for the appearance of action, rather than demand action itself. Reform will threaten every one of us—and it should.

What will reform require? I mentioned the main goals at the start, so let me briefly outline the problems as I see them.

- Our students are undereducated. National tests and international comparisons find our precollege students poorly informed in science, mathematics, geography, and whatever else the testers choose to test. Some students in Texas are unsure of what lies south of the Rio Grande—that is, if they know about the Rio Grande. Others cannot place the United States on a blank map of the world. We hope to educate these young souls so that our nation can remain a world leader, but this may be difficult if they are not all that sure where the world is.
- And why are they ignorant? It cannot be a deterioration of their genetic makeup, so it must be a combination of how they are raised and how they are taught. We can do little about how they are raised, but we must do much more about how they are taught—at all levels. How do the teachers stack up? Some are surely among the most wonderful, dedicated, and competent members of society. At the same time, many are poorly trained in the sciences, and many high-school biology classes are taught by former majors in home economics or physical education. Salaries and working conditions for teachers are often such that few of the most gifted undergraduates would consider a teaching career in the precollege grades.

- Should we blame the teachers? No; and now we reach the crux of the problem—we should blame the colleges and universities. It is they which select, educate, and certify the teachers-to-be. If there is something wrong with the teachers' education, the universities cannot escape a major responsibility. (I am excluding from analysis all the other factors, about which we can do little, that tend to lessen the effectiveness of teachers: salary, working conditions, position in society, etc.).

How do we scholars in the great universities—the flagships of education, so to speak—go about educating the teachers-to-be for what is one of the most critical tasks in society? For the most part, we ignore these young students, rarely encourage them to undertake what should be a noble career, and at times actively discourage them by suggesting that they will be wasting their lives.

I know of no disciplinary department in a great university that would consider it acceptable to encourage and help to educate an outstanding student for a career of teaching in the schools. The goal of the education we profess has Stockholm, not the Little Red School House, at the end of the road.

And why does this (to me) intolerable situation exist? The answer from the typical university professor in science is that one can get away with such behavior, and in fact there is strong encouragement to do so. The criterion for advancement and reputation is research, but even that is being replaced by the size of one's research grant.

Gone are the days when fine scholarship *and* fine teaching were demanded by the system. The view now seems to be that any fineness devoted to teaching must mean less fineness in research and grantsmanship. In a zero-sum game, it cannot be otherwise.

This state of affairs exists because those who lead the universities and those who lead society make little effort to promote or demand a deep commitment to quality education. In fact, one can maintain that there is no national leadership in education. The educational system costs more in money and manpower than does the Pentagon—yet we have no generals. The educational system works on the principle of letting 1,000 flowers bloom, but too many of those flowers have withered.

Let me give a specific example of how one segment of leadership works—or does not work. A mandate of the National Science Foundation (NSF) and the National Institutes of Health is to support science. This is normally interpreted by the agencies and the scientific establishment to mean support of scientific research. One most promising way to do this is to support the work of first-class scientists. An equally important activity would be to support first-class education for the K-16 years. A generation ago, NSF did gallant and effective work in science education, but then reaction to "Man, a Course of Study" (MACOS) and pressures from far-right politicians soured the educational programs. But don't blame NSF entirely. When all this was going on, both the scientific and educational establishments tended to look the other way. Only recently has NSF again shown a deep interest in educational reform. Let us hope that a balance will be reached and that the agencies will be equally concerned with the production of good science and of good scientists.

NSF's basic concern with the support of research has a negative effect on education. Consider this: One can include in a research proposal requests for funds to hire someone to do the teaching of the principal investigator, so the principal investigator can get on with the real work—research. Could there be a clearer message of the relative importance of teaching and research? How bankrupt can we get?

This attitude pervades the administrations and faculties of the universities. Excellence in research confers status and prestige for both the individual and the institution, but must that result in the acceptance of less-distinguished or even mediocre teaching? The futures of science, society, and the universities all depend on quality education. If we forget that, surely we are shooting ourselves in the foot.

But what can be done? All that needs to be, if we so wish. Nothing new has to be discovered, should we decide that education must be reformed; furthermore, a great deal could be accomplished even with existing resources. Let us assume that our goal really is to prepare our students to be able to make informed decisions about themselves, their communities, their nation, and their world in the on-rushing serious problems that have a scientific component.

The first thing that we have to accept is that the minuscule amount of science that our students receive is inadequate. We have been told that there is only a trivial difference in knowledge of biology between students who have taken a biology course and those who have not.

I suggest that our goal can be achieved only if the percentage of school work devoted to the sciences is increased to about 20-25% of the curriculum, instead of the 2-5% that now prevails. And I am most certainly not talking about a 10-fold increase in the sort of science now generally taught. We all know the sort needed: hands-on activities, inquiry orientation, interdisciplinary approaches (including the nonscience disciplines), emphasis on concepts, and the avoidance of a plethora of unneeded and unused facts.

Most important, science education must begin with young children, and they must be partners with teachers who will help them uncover nature and not stifle that sense of wonder and joy in living things that is every child's birthright.

The science taught must be organized to accomplish the most with a minimum of repetition. If the elementary-school years could be devoted to learning about nature—animals, plants, and the environment—and the middle-school years to ourselves and our place in that environment, the young student would reach the high-school years with an understanding of biology considerably better than that held by students who have had only a year of tenth-grade biology.

That would mean that high-school biology could be set at a level that would allow serious and informed consideration, not only of important human problems that biology and the other sciences have something to say about, but also of selected areas of modern biology that are providing such deep and exciting insights into the phenomena of the natural world.

Such a plan would make meaningless that recurring question: How can we cover biology in a single year, when we have to include all those current major advances in this major field of science? It is astonishing that we ever thought we could, or should.

All this will require a revolution in the way biology and the other sciences are taught, and such a revolution must involve the entire nation; and above all it demands direction and leadership. I suggest that such leadership must come from a highly respected, nongovernment organization that will support reform of the teaching of science at all levels: seeking to increase the fraction of the curriculum devoted to science, improving the teaching force, demanding adequate resources from local and national governments, seeing that fine textbooks are available, encouraging the colleges and universities to take seriously the education of all students (including those seeking a career in teaching), and providing models for appropriate science for the various grades. One of the most important functions for such an organization would be to make suggestions for a sequence of topics appropriate for each grade level. There must be more science in the curriculum, and there must be nationwide agreement on what this should be.

If we accept that far more science must be taught, a common weakness in our standard approach will be avoided. Whenever we assemble to talk about what should be taught in high-school biology, we make our recommendations on the assumptions that students have had no biology before and that we need not lay a basis for any that might be taken in more advanced grades. The goal should be to establish what biology is to be taught in each grade, not which grade is to receive the single massive dose.

Another thing that a central organization could do, in contrast with what most committees do, is recognize that many people throughout many years have been dealing with the same basic problems. There is a body of information and experience that is valuable and should not be ignored by each new committee—as it often is. A central organization could not only synthesize what is known, but also coordinate the efforts of all toward the improvement of science teaching in the schools.

Such an organization should be headed by the most vigorous and visible leaders of our great universities, scientific institutions, and public and government organizations. I am talking about powerful people who care and who are effective in making a clear distinction between activity and accomplishment. Such leadership could supervise the permanent staff and associated committees that would attempt to translate goals into programs and products. Such an activity would have to involve real teachers and close working relations with state and local school districts, publishers, and the institutions that prepare students for careers in teaching.

Will society, and especially the educational establishment, buy this? Probably not, without considerable pressure. But is there any alternative, if we are to achieve our stated goal? There must be a national group that will set standards and offer advice. There cannot be a nationwide reform unless there is an organized nationwide reform effort. Surely we have enough evidence to recognize the nearly total ineffectiveness of this seemingly endless stream of committees bent on educational reform that merely promulgate, and then disband. There is no longer a need to analyze what is wrong; we know what it is and in a general way what the remedies must be. In fact, there may be merit in proposing a moratorium on reform-minded committees unless there is a firm link to a planned program or product. Maybe we can help to make something very worthwhile happen. We had better. We recognize that our educational system is short-changing the nation and that the system is our responsibility. But even more serious is the fact that we are short-changing our young people and those who teach them.

We must change the system so that students will understand and take joy in the natural world and protect it, as it in turn provides for them. They must be able to deal with the many serious problems that affect all of us and wild nature as well. And we must change the system so that teachers can take joy in their profession and what they profess and will be allowed to hold the position that they should in society—because they are doing the basic work of civilization.

Sustainable reform of high-school biology will require far more than tinkering with the high-school biology curriculum. That approach has been tried repeatedly, and the problem is, if anything, more serious than ever. Sustainable reform will come only when the colleges and universities effectively educate those who will teach in the schools. The high-school science-curriculum course must be a culmination of the students' rich experience in biology and the other sciences throughout elementary-school and junior-high-school years.

We must also explore mechanisms for developing greatly improved text and laboratory materials.

This would be a radical reform and realistically would require implementation over a period of years. Thus, an interim solution would seem to require two sorts of high-school programs. The first, and transitory, course would take the students as they are now and provide a single-year course as good as possible—so good that the students would know more biology at the end than those who had not taken the course.

The development of a second type of high-school course should become the main thrust of our efforts. It would be the culminating and synthetic approach mentioned before and would be based on a good knowledge of science obtained in the elementary-school and middle-school years. It could be taken in the tenth grade. If this ideal K-10 program could be achieved, it is more than likely that many students would profit from a more advanced course in the twelfth grade.

To make all this possible, we should explore the possibility of encouraging the formation of a new, permanent, nationwide organization or the modification and energizing of an existing one, to catalyze the reform.

> In the conditions of modern life . . . the race which does not value trained intelligence is doomed. Not all your heroism, not all your social charm, not all your victories on land or at sea, can move back the finger of fate. To-day we maintain ourselves. To-morrow science will have moved forward yet one more step, and there will be no appeal from the judgment which will then be pronounced on the uneducated. [Whitehead, 1929].

REFERENCES

Koshland, D. E., Jr. 1988. Thinking tough. Science 241:1273.
Whitehead, A. N. 1929. The Aims of Education, pp. 22-23. New York: Macmillan.

28

Institutional Barriers to School Change

PETER W. AIRASIAN

INTRODUCTION

Identifying institutional barriers to meaningful educational change requires consideration of schools at two levels. First, schools as a group must be viewed as social institutions that interact with and are influenced by an array of other social institutions. Schools are not free to operate independently of these external social agencies and institutions, which look to the schools to foster a variety of desired personal and social outcomes in pupils. Second, schools and school systems must be viewed as entities unto themselves, each with its own bureaucracy, personnel, budget, clients, and resources. The dynamics existing within and among these bureaucratic factors create inherent barriers to change.

At both levels, schools are best thought of as conservative institutions; their inherent impetus for change is not great, and their programs, policies, goals, and agendas are determined largely by groups external to the schools or the school system (Cremin, 1961; Fullan, 1982; Nyberg and Egan, 1981). Schools do change, but changes typically are imposed by external institutions or groups. If we consider significant, large-scale educational reform movements of the last quarter century—such as mandated state

Peter W. Airasian received an A.B. in chemistry in 1964 from Harvard College and has taught high-school chemistry and biology. He received an A.M. and Ph.D. in educational testing and evaluation from the University of Chicago. He is professor and chair of the Educational Foundations Division, Boston College.

testing programs, school finance reform, teacher promotion ladders, and opening of schools to a variety of special-needs pupils—it is clear that the reforms were initiated, championed, and eventually enacted by groups external to the schools, usually state-level elected officials, businessmen, and the courts. Even though the reforms influenced important aspects of teachers' and school administrators' activities, the reforms did not originate in the schools or educational community. In fact, many educators opposed these reforms when they were proposed initially. Historically, educators have been charged with implementing reform programs that they have had little influence in creating or enacting.

The purpose of this discussion is to consider some of the external factors that influence the nature of social mandates for school change and the internal realities that place limits on the responses to these mandates that schools can muster. Because it focuses on barriers to change, this presentation may be perceived to be pessimistic. The intention is not pessimism, but realism. It is hoped that the discussion will counterbalance the promises of reformers who inevitably will be enthusiastically optimistic and exclaim broadly about the many beneficial outcomes of their reform proposals.

The bulk of the paper is concerned with general barriers to large-scale school change; the discussion is not focused on a particular curriculum area or grade level. First, I describe the genesis of change movements and how they are shaped and influenced by the status of schools and schooling. The consequences of this status for large-scale, mandated school change are considered. Then I describe internal, school-based factors that inhibit change. The factors discussed represent a view of the American educational scene that points out the difficulty of effecting real educational change without a substantial commitment of resources and a substantial amount of patience.

EXTERNAL BARRIERS TO SCHOOL CHANGE

Change efforts arise when a crisis is perceived to exist in a social agency; reform is not spontaneous, but responds to a perceived need. For example, the reform of science curricula in the 1960s and the more recent push for higher academic standards in schools were the result of the unexpected launch of Sputnik and 10 years of publicity about the poor test performance of American schoolchildren, respectively. It is not clear how such crises emerge, why some capture the attention of the public while others do not, and why attention typically shifts to another crisis after a relatively short period. What is clear, however, is that a prerequisite for large-scale reform is a public sense of urgency about a social or educational problem.

It is important to note that the nature of change needed to ameliorate a perceived problem is not necessarily evident from the problem itself. When pressed to "do something now," policy-makers have a choice of many actions that may be taken. A decline in the levels of academic achievement among schoolchildren can be attacked by extending the length of the schoolday, adding required courses to the school curriculum, mandating exit tests to certify pupils' competence, instituting remedial programs, writing new textbooks, raising grade-to-grade promotion requirements, increasing funding for schools, or some combination of these strategies. Even this array of potential reform strategies is incomplete, since none considers the role of nonschool sources, such as the family, in solving the declining achievement problem. The factors thought to underlie the problem will have much to do with the strategies selected to correct matters. Different perceptions of underlying causes evoke different types of reform efforts. Hawley (1985, p. 184) correctly states that

> the policy implications of the premise that our schools have fallen from some former state of grace are very different from those one would derive from the belief that our schools, overall, are as good as they have been but that this is not good enough for the challenges we now face. In the first case, the solution lies in shaping up the schools, cutting out incompetence, and making incremental improvements. The view that we need to do better than our previous best, on the other hand, dictates that we worry less about defining limits and setting standards and more about investing in new capabilities and reshaping basic ways we go about the facilitation of student learning.

Thus, the impetus for reform derives from the emergence of a public perception that something is awry in some segment of society. Without such a perception it will be difficult to rally public support for reform. The perceptions of the policy-makers regarding the underlying causes of the problem narrow, in turn, the nature and locus of reform efforts. Once a problem is identified and school-based change is perceived to be needed, the nature of the proposed change mandate will be forged within the realities of the present status of schools as a social institution. In this regard, the last 25 years have seen a number of factors produce changes in the status, practices, and priorities of the educational system that influence and present barriers to meaningful school reform (Airasian, 1987). Three such changes are growth, centralization, and politicization.

Growth

On virtually any index other than the numbers of pupils and teachers, there has been growth in the educational system over the last 20 years. Expenditures for education have soared. While the Consumer Price Index tripled between 1960 and 1982, the expenditures for public elementary-school and secondary-school education increased sevenfold. The educational system has also grown in the diversity of the pupil groups it serves and in the variety of programs it offers these groups. Twenty years ago, the vast majority of school pupils would have been undifferentiated, except for their reading-group placement or high-school track. Today, as a result of legislative and judicial mandates, schools provide varied programs, services, and tracks to many pupils who would not even have been allowed to enroll in public schools 2 decades ago. The school is a much more finely differentiated institution now than in the recent past.

Along with growth in expenditures and pupil differentiation, there has been substantial growth in the number of goals and functions that the schools are expected to attend to and attain. This growth is a measure of the increasing importance of the schools as social initiation agencies and the apparent inability or unwillingness of other social institutions to take responsibility for fostering the desired outcomes.

The American public desires that its schools be responsible for a large set of personal and social goals and activities that once were considered the province of other institutions. Thus, at the personal level, schools are expected to outfit pupils with a growing array of social and "survival" skills: education as to sex, AIDS, drug and alcohol abuse, conflict resolution, and values clarification; practical mathematics; health and nutrition; race relations; training for parenthood; English for daily living; and check-writing, map-reading, and the like. There are calls for such science courses as "Chemistry for the Consumer," "Science and Society," and "The Biology of Being Human."

The curriculum has expanded, primarily through a multitude of "functional-skills" courses that are designed to make pupils' lives in society possible, if not worth while. Each new course and requirement competes for time and attention with other "functional-skills" courses, as well as courses in more traditional academic disciplines. One study (Goldenstein et al., 1988) indicated that the proportion of courses that high-school pupils took in the areas of English, social studies, business, mathematics, and natural sciences declined from 76% in 1953 to 64% in 1983. Schools are expected to help pupils to attain both the new socialization goals and more traditional academic subject-matter mastery in a manner that satisfies the two more general social goals of equality and excellence. This is not an easy task.

CENTRALIZATION

As the education system has grown in terms of expenditures and responsibilities, the interest of the state in education also has expanded

(Green, 1980; Wirt and Kirst, 1982). As education becomes more narrow and instrumental and concerned with "basic," "functional," and "survival" skills, it becomes less a privilege and more a requirement for life—a fact that increases the interest and stake of the state in the process and outcomes of education.

Moreover, the increase in mandated and legislated state and federal education programs has produced increased government intervention and control intended to maintain uniformity and fairness of and control over an increasingly diverse array of local educational activities and responsibilities (Wise, 1979). The sheer weight of mandates and laws plus the social, political, and legal consequences of failure to carry them out properly has prompted centralization and standardization of educational decision-making. Standardization also works from the bottom up, as evidenced by specific job definitions, regulations, and expectations that are negotiated into many teachers-union contracts. While we are beginning to see a glimmer of interest in returning control of education to persons at the local building level, the predominant political belief is that the job of education in the 1980s is too big and complex to be left primarily in the hands of the family or local school district.

Increasingly, criteria for and decisions about curriculum, instruction, certification, funding, evaluation, and desired outcomes are being made at the state or national level, not the local level. Decisions are made on a cross-unit basis, rather than a within-unit basis, with the consequence of reduced discretion at the local school-building level. The values advanced to justify increased state control—namely, equality, accountability, efficiency—clash with the values of freedom of choice and differentiated treatment, which are embodied in justifications for local school control. Another important consequence of centralized decision-making is that state governments are becoming more involved in the details of teaching, testing, and curriculum than ever before. In many instances, policy-making is being given over to less expert, more politically motivated groups that are as concerned with how decisions will play in Peoria as they are with how much real educational change will be produced. Individuals who wish to bring about change in the schools must recognize and contend with this movement toward administrative centralization and standardization.

POLITICIZATION

A third change in schools in the last 25 years is that education and the educational decision-making process have become overtly politicized (Hansot and Tyack, 1982). Schools always have been confronted by a large number of competing agendas, so the problem of accommodating political diversity is not a new one for education policy-makers. What is new, however, is the increased number of politically active, legally and legislatively protected special-interest groups spawned by events of the last 2 decades (Murphy, 1982).

Each group has its agenda for the schools, its turf to protect, and a stake in the decisions that affect educational practices and priorities. Different value premises are implicit in the agendas of different groups, as can be seen in the varied answers that one receives to fundamental questions, such as: "What attributes ought to determine the distribution of educational resources to pupils?" "Should academic, social, or personal goals take precedence in the schools, and which goals within the selected categories?" Administrators and legislators at all levels seek to answer the question: "How can I allocate limited funds, resources, curriculum time, rights, and prerogatives to an ever more differentiated school population in the fairest manner possible?" They know that any policy that does not provide more for every group and interest will be hotly contested, but that any policy that does produce more for everybody will be self-defeating vis-à-vis social goals, such as equality.

Prospective reformers of schooling must recognize the context into which they venture under the banner of educational change. In the late 1980s, compared with earlier times, schools are charged with serving a very heterogeneous mix of pupils in a broader range of ways on a greater set of anticipated outcomes with quality and without discrimination in the face of increased public scrutiny, value diversity, and suspicion.

IMPLICATIONS

The growth, centralization, and politicization inherent in the educational system have consequences for the nature of the school-change mandates that emerge from external groups and institutions. In looking at recent reform efforts, we can see a number of common characteristics.

First, as noted, the reforms were initiated, championed, and eventually enacted through the efforts of the courts, state-level elected officials, and the business community.

Second, as might be expected, given the pessimism about American education that has prevailed of late and the need to administer reforms at a statewide level, reforms tended to have two characteristics: they focused on minimum levels of expectation and they were relatively easily implemented and managed. With very few exceptions, the state has been interested in and preoccupied with providing an education that is suitable "on average" for all school pupils. Developing individual pupils to their maximum level of attainment has not been a usual priority of the state in formulating educational policy (Green, 1980); hence, the focus in recent reforms has been on minimally acceptable levels of funding, standards, and attainment.

Similarly, the reforms for the most part were easily implemented and managed, calling for the administration of standardized tests, the enactment of new required-course-load policies, and the application of new revenue-distribution formulas. There was little attempt to intrude directly into classrooms, except to use such reforms as mandated tests to push the curriculum in desired directions. Few reforms were concerned with how things were taught or, in most cases, what specific topics were taught.

Third, reflecting the politicized nature of education issues, the reforms were politically palatable to varied education constituencies. The language of reform emphasized terms like "basic skills," "minimum competence," "higher standards," "rewards for good performance," and "maintaining the country's competitive edge" to strike a responsive chord in large segments of the public, who were ready to support such general principles. Blue-ribbon panels, state legislatures, and national commissions deliberated over the problems of schools and typically came to conclusions that exaggerated problems; produced very general recommendations focused on ends, not means; and protected existing institutional arrangements by only rarely calling for basic institutional change. The fact that many of the reforms either were not spelled out in detail or involved an activity peripheral to what went on from day to day in classrooms (e.g., high-school graduation testing) deflected controversy that might have emerged if the reforms had been more intrusive or clearly delineated. The language of a reform and the degree to which it finds a nonthreatening position on the political spectrum are potent factors in determining its acceptability.

Fourth, the press for statewide, minimum, politically acceptable reforms has produced many reforms that are symbolic of desirable social and educational values, rather than directive of classroom activities and outcomes (Airasian, 1988). The reforms tend to ignore the complexity of school and educational improvement by offering simplistic, one-variable solutions and bromides. The rhetoric and the symbolism of reform conspire to produce two potential barriers to school change: the grandiose claims reflected in the language of reform lead to raised and often unrealistic expectations of what schools can and should accomplish, particularly since recent reforms place scant emphasis on the schools' working in conjunction with other institutions, such as the family; and the political and symbolic appeal of the reforms often serves to assuage public concern or conscience about the need for further reform. Public attention to and concern about the existence of a problem are prerequisites of externally imposed school reform—perceived crisis provokes responses. But the public has a short attention span and flits from social problem to social problem at rather short intervals (Airasian, 1983). The danger of reforms that are largely symbolic or that do not attack root causes is that they can become conscience-salving solutions to what are real problems—solutions that provide psychic satisfaction to policy-makers and the public, but little in the way of meaningful reform in school and student outcomes.

Finally, most educational reforms are adopted with high levels of technical uncertainty; the wisdom of their adoption and the range of their effects rarely are known in advance of implementation. A variety of political motives dictate that reforms not be examined empirically before their adoption, so justification for reforms is found most often in their "social and political validity" (Airasian, 1988), not in hard-headed evaluation of their outcomes and consequences. Logic and rational planning are all too often upset when they confront the exigencies of the real world, and George Bernard Shaw was correct when he pointed out that one of the common errors made by social reformers was the idea that change could be achieved by "brute sanity."

INTERNAL BARRIERS TO SCHOOL CHANGE

The preceding context colors substantially the reforms that emerge for implementation in schools. The reality of the school as a social institution that takes its lead from other social institutions is that change is most likely to occur where there is widespread public awareness and concern; where basic, minimal expectations can be stated; and where interest groups can find a common, usually very general ground for proposed change.

Reforms that meet these criteria have been of two types. The first type calls for a change in some policy, regulation, or administrative practice; the second calls for a change in the outcomes of schooling. Examples of the first type include requiring all pupils to complete three science courses in order to receive a high-school diploma, mandating new high-school graduation and teacher-certification testing programs to serve as gatekeeping mechanisms, adding access ramps to public buildings and walkways, and applying new funding formulas to alter the distribution of school resources. Many of the recent mandates to reform education have been of this first type and involve management of the school environment and instructional delivery system. The schools respond to such mandates by adding a new course, accommodating a statewide test once or twice a year, replacing stairs with ramps, and dividing resources in new ways. The second type of change mandate calls for a change in the outcomes of education or in the way that those outcomes are reached. This type of mandate is concerned with issues of what is taught, how it is taught, and whether students attain desired competence. Thus, when critics rail against the "dry and uninteresting" science textbooks that stress "recall of abstract facts" and seek to substitute science education that is "geared to the interests of the pupils and focused on their ability to think rationally and grapple with and solve problems," they are asking for more than a change in administrative or bureaucratic regulations. They are calling for a fundamental change in the way science education is delivered to pupils.

The two types of reform are similar insofar as the likelihood of their success is unknown at the time they are implemented. They are different, however, in the way they are phrased and in the guidance they offer school people to carry out their implementation (Elmore and McLaughlin, 1982). Policies of the administrative-managerial type generally contain the knowledge necessary to implement them. Policies of the second, instructional type often do not contain the knowledge necessary to implement them. Consider the difference between the administrative-managerial mandate to "put ramps at the entrance to all school buildings" and the instructional mandate to "improve the educational achievement of all handicapped pupils" or the difference between "establish a required three-course science sequence that all high-school pupils must take" and "teach high-school pupils to reason logically about scientific issues." We know what to do to meet the administrative-managerial mandates. It is considerably less obvious what needs to be done to meet the instructional-outcome mandates (Timar and Kirp, 1988).

It is the latter, instructionally focused type of school-change mandate that requires particular consideration, because as administratively based reforms have been implemented in schools, reformers have begun to expand their activities by seeking to redefine and extend the outcomes that pupils are expected to attain in school. This latter type of change has an impact directly on the instructional process and depends on the state of current knowledge about instruction for its success. The main focus of the following discussion is on the current level of our instructional expectations and the degree to which our mastery of the intricacies of the instructional process are sufficient to meet these new expectations.

It is necessary first to note briefly some other important internal barriers to school change (Airasian, 1983; Fullan, 1982; Lieberman, 1982):

- There is the reality of the inherent nature of all bureaucracy to resist innovations that change the resource-allocation mix or accustomed authority relationships. This is particularly true in schools, where a change usually means adding something to role or curriculum responsibilities without reducing those responsibilities in some other arena. Reforms must be capable of being accommodated into an already hectic, fragmented, crowded school calendar.
- There is the reality of teachers carrying forward the strengths and weaknesses of their own training into the classrooms where they teach. Schools are staffed by former school pupils. The deficiencies of education

for one generation often get carried forth into the instruction of succeeding generations of pupils.

- There is the reality that learning—both cognitive and, perhaps more important, affective—is cumulative in many school subjects. Because of this fact, it is often difficult to intervene at later stages of the school sequence to rectify inadequacies developed at earlier stages.
- There is a reality of the different world views of reformers and classroom teachers that hampers implementation and teacher commitment to school reform. Reformers live and work in a world that is formal, precise, prescriptive, concerned with ideas, and dedicated to belief in the ideas that they offer for reform. Teachers live and work in a world that is informal, loose, experimental, concerned with process, and dedicated to adaptation to the classroom reality.
- Finally, there is the reality of limits to the amount of meaningful change that schools can accomplish without the support, reinforcement, and encouragement of families and other social institutions.

A major additional barrier to instructionally related educational change is the lack of a well-developed science of instruction that will permit teachers to attain the many and varied instructional ends that society wishes to have schools foster in pupils. Recent years have seen an increased call for schools to emphasize thinking, reasoning, and problem-solving skills in instruction, in place of more rote learning in various curriculum areas, including science. The vision of such outcomes has led many reform advocates to underestimate the instructional complexities involved in attaining these nonrote ends and to view the issue of instructing pupils in nonrote behaviors as a mere technical problem readily solvable by application of existing, standard procedures (Wise, 1979; Raywid, 1986).

Increasingly, reform mandates, decisions, practices, and desired student outcomes outstrip the evidence available to warrant them (Wise, 1979). Most instructionally based change programs are predicated on the implicit assumption that the science of instruction is sufficiently developed that teachers can apply it to guide pupils to the attainment of virtually any desired instructional outcome. Stated in a more down-to-earth manner, the assumption is that we know how to teach pupils the many and varied outcomes that we have stated in our reform programs. This is a critical assumption, for if proposed reforms are to lead to the attainment of new student outcomes, it is necessary that available instructional knowledge and techniques be sufficient to enable teachers to teach most pupils the desired outcomes.

The distinction between wishes and expectations has been lost or ignored in much of what has passed in the last decade and continues to pass for instructional reform (Green, 1980). We have let our wishes become our expectations, while paying scant attention to what is known and not known about the educational and instructional processes.

When the content focus of instruction is on mastery of general, minimum, lower-level skills and facts, instructional knowledge is not a major barrier to successful instruction and learning; enough is known about appropriate instructional techniques to assume that most teachers could foster the desired rote skills and knowledge in their pupils (Fredericksen, 1984). Recently, however, many reformers have sought to up the instructional ante by mandating school programs that seek pupil mastery of more cognitively complex objectives. Discussion of curriculum reform in science and other disciplines calls for a reduction in the amount of "cookbook fact acquisition" with a corresponding increase in such outcomes as thinking rationally, solving problems, applying known information to solve unfamiliar problems, and analyzing situations through logical or critical reasoning. Such higher-level cognitive pupil behaviors are a far cry from recalling facts and memorizing formulas. The hallmark of higher-level cognitive behaviors is the use of mental processes more complex than rote memorization to obtain the solution to new and unfamiliar problems (Bloom et al., 1956). Thus, the two defining properties of higher-level processes are that they involve problems that require more than rote learning and are new to the learner. Researchers who have studied higher-level mental operations like critical thinking, problem-solving, and reasoning have obtained results that are not encouraging for reformers who wish to reorient instruction to produce such outcomes in pupils.

Various reviews indicate that teaching higher-level behaviors is different in many ways from teaching lower-level, rote behaviors (Cuban, 1984; Derry and Murphy, 1986; Fredericksen, 1984; Wade, 1983). Higher-level behaviors take longer to teach; they develop gradually over time, not via the quick association and closure that often occurs in teaching rote material. Direct instruction is useful for teaching lower-level tasks and knowledge, but higher-level behaviors call for less direct, less structured instructional environments and methods. Acquiring higher-level processes requires that pupils be provided opportunities for self-discovery. The content of higher-level processes may be precisely the type of content that is difficult to localize in the curriculum and too diffuse to focus instruction. The present organizational and instructional arrangement of schools may make them inhospitable environments for fostering reasoning and other higher-level behaviors.

When instruction can be reduced to teaching pupils an algorithm to apply in a specific situation and giving them a great deal of practice in applying the algorithm, learning generally takes place. After they learn the steps involved in solving simultaneous equations and after they practice by solving 100 simultaneous equations, we anticipate that students will "apply" their knowledge when confronted with a "new" simultaneous-equation problem of a structure similar to that of previous problems. However, if the problem cannot be expressed in terms of an algorithm or if it is not narrowly confined in content and format, evidence suggests that we know little about useful instructional strategies to instill problem-solving capabilities (Fredericksen, 1984; Derry and Murphy, 1986).

> But as we go into domains where problems are increasingly ill-structured, we can be much less certain about the adequacy of our knowledge. We know little about how to teach students to develop representations of ill-structured problems, to develop plans for solving these problems, or to employ appropriate strategies or heuristic approaches. Still less can we advise students about efficient methods for accessing relevant information in [long-term memory] [Fredericksen, 1984, p. 396].

No clear body of knowledge exists regarding the conceptualization of higher-level behaviors, such as reasoning, logical and critical thinking, and application; nor are there well-validated instructional strategies to teach such higher-level processes:

> How do you teach daily 25 to 35 students compelled to attend class an indeterminate reasoning process that, in order to be learned, must be individually understood, applied and assessed indirectly through each student's words and behavior while teaching a group in a limited amount of time? No answer exists for this question. No technology of teaching produces reliably (don't even consider validity) acceptable outcomes in reasoning [Cuban, 1984, p. 672].

Once desired objectives venture beyond application of an algorithm in a narrowly defined problem situation, the outlook for validated instructional techniques that are capable of fostering more general, abstract higher-level behaviors in most pupils is bleak. At present, processes like reasoning, critical thinking, problem-solving, and the like are beyond our capabilities to instruct in a manner that can ensure that most pupils will master them. The moral here is that we cannot assume that, because we can state the desired outcomes that we wish pupils to attain from instruction, it necessarily follows that the science of instruction is sufficiently well developed and articulated to enable us to be successful in fostering them in pupils. Note also that the same argument can be made for reforms that focus on teachers and seek to make them more dedicated to their subject matter and more interested in their students. Too often proposed instructional change has foundered, because the expectations of the reformers have overlooked or far exceeded the realities and limitations of classrooms, teachers, and instructional knowledge.

SUMMARY

This paper asked and tried to answer three questions that influence the nature of change in educational institutions: What is the role and current perception of schools in American society? What is the nature of educational-reform mandates that emerge from external sources to solve what are perceived to be school-based problems? What are the internal realities of schools that reformers will encounter in seeking to implement their mandates? The answers to these questions indicate that barriers to educational change exist at many levels. It has been noted that the bureaucratization and politicization of schools have created a number of barriers to accomplishing meaningful change in schools. Among these barriers are the focus on minimally acceptable changes, the need for politically palatable reforms, the often symbolic nature of reforms, and the lack of testing of reforms before widespread implementation. Within the school, where implementation of externally imposed reforms takes place, the main barrier to meaningful changes in the outcomes of education has been the choice of unrealistic desired outcomes in the face of limited knowledge of how to foster such outcomes in the classroom.

It is not likely that the barriers noted above will disappear in the near future; they will continue to be a reality of the educational landscape that reformers will have to confront. The barriers should not deter us from trying to accomplish change, but they should make us realistic about our efforts. Grandiose, Band-Aid efforts will not suffice, except to provide temporary political capital. Meaningful change will come about when we set our aims reasonably in the light of existing barriers; when we can be patient about the pace at which change will occur; when we stop relying on the school, independently of other social institutions, to solve problems that are not fundamentally school-based; and when reformers are better versed in the realities of schools and classrooms.

ACKNOWLEDGMENT

The insights and suggestions of Dr. Vincent C. Nuccio added greatly to the preparation of this paper. The sole responsibility for the final content of the paper, however, rests with the author.

REFERENCES

Airasian, P. W. 1983. Societal experimentation, pp. 163-176. In G. Madaus, M. Scriven, and D. Stufflebeam, Ed. Evaluation Models. Boston: Kluwer-Nijhoff.

Airasian, P. W. 1987. State mandated testing and educational reform: Context and consequences. Amer. J. Educ. 95:393-412.

Airasian, P. W. 1988. Symbolic validation: The case of state-mandated, high-stakes testing. Educ. Eval. Pol. Anal. 10:301-313.

Bloom, B. S., M. D. Engelhart, E. J. Furst, W. H. Hill, and D. R. Krathwohl. 1956. Taxonomy of Educational Objectives. Handbook I: Cognitive Domain. New York: David McKay Co.

Cremin, L. A. 1961. The Transformation of the School. New York: Alfred A. Knopf.

Cuban, L. 1984. Policy and research dilemmas in the teaching of reasoning: Unplanned designs. Rev. Educ. Res. 54:655-681.

Derry, S. J., and D. A. Murphy. 1986. Designing systems that train learning ability: From theory to practice. Rev. Educ. Res. 56:1-39.

Elmore, R. F., and M. W. McLaughlin. 1982. Strategic choice in federal education policy: The compliance-assistance trade off, pp. 159-194. In A. Lieberman and M. McLaughlin, Eds. Policy Making in Education. Eighty-first Yearbook of the National Society for the Study of Education. Part I. Chicago: University of Chicago Press.

Fredericksen, N. 1984. Implications of cognitive theory for instruction in problem solving. Rev. Educ. Res. 54:367-407.

Fullan, M. 1982. The Meaning of Educational Change. New York: Teachers College Press.

Goldenstein, E. H., R. R. Ronning, and L. J. Walter. 1988. Course selection across three decades as a measure of curriculum change. Educ. Leader. 46:56-59.

Green, T. F. 1980. Predicting the Behavior of the Educational System. Syracuse, N.Y.: Syracuse University Press.

Hansot, E., and D. Tyack. 1982. A usable past: Using history in educational policy, pp. 1-22. In A. Lieberman and M. McLaughlin, Eds. Policy Making in Education. Eighty-first Yearbook of the National Society for the Study of Education. Part I. Chicago: University of Chicago Press.

Hawley, W. D. 1985. False premises, false promises: The mythical character of public discourse about education. Phi Delta Kappan 67:183-187.

Lieberman, A. 1982. Practice makes policy: The tensions of school improvement, pp. 249-270. In A. Lieberman and M. McLaughlin, Eds. Policy Making in Education. Eighty-first Yearbook of the National Society for the Study of Education. Part I. Chicago: University of Chicago Press.

Murphy, J. 1982. Progress and problems: The paradox of state reform, pp. 195-214. In A. Lieberman and M. McLaughlin, Eds. Policy Making in Education. Eighty-first Yearbook of the National Society for the Study of Education. Part I. Chicago: University of Chicago Press.

Nyberg, D., and K. Egan. 1981. The Erosion of Education—Socialization and the Schools. New York: Teachers College Press.

Raywid, M. A. 1986. Some moral dimensions of administrative theory and practice. Iss. Educ. 4:151-166.

Timar, T. B., and D. L. Kirp. 1988. State efforts to reform schools: Treading between a regulatory swamp and an English garden. Educ. Eval. Pol. Anal. 10:75-88.

Wade, S. E. 1983. A synthesis of the research for improving reading in the social studies. Rev. Educ. Res. 53:461-497.

Wirt, F. M., and M. W. Kirst. 1982. Schools in Conflict. Berkeley, Calif.: McCutcheon Publishing Co.

Wise, A. E. 1979. Legislated Learning. Los Angeles: University of California Press.

29

State Policy Tools for Educational Reform—Barriers or Levers for Change?

JANE ARMSTRONG

Barriers to science education are not endemic in biology or science education alone. Recent results from the Nation's Report Card (Applebee et al., 1989) indicate that student achievement is unacceptably low across all subject areas—not just in science. This suggests serious problems and barriers in our educational system, not just in how we teach science.

To a large degree, the structures and functions of our educational system are rooted in state policies, mandates, and regulations. State policies have both helped and unintentionally hindered educational reform. Although recent reforms have raised standards of educational quality, they have also set in place some regulations that hamper the systemwide change that is now needed to improve student learning.

It is because 5 years of reform at the state level has gotten us where we are now that I want to describe these reforms, highlight some positive impacts, and suggest new strategies to revise existing policies or develop new policies that will help to improve our educational system, including biology education.

This paper is organized around four themes:

Jane Armstrong is a senior policy analyst at the Education Commission of the States in Denver. Dr. Armstrong has a background in biology and in educational research and measurement. She recently completed a study funded by the National Science Foundation on the impact of state policies on science curriculum. She has served as an advisor to the First International Mathematics Study and on the editorial board of the *Journal for Research in Mathematics Education*.

- First, I will describe where existing reform and state policy have gotten us, in terms of classroom instruction.
- Second, results are presented from a study recently completed at the Education Commission of the States (ECS) that highlights the impacts of recent state reforms on science education in school districts and schools.
- Third, a new reform agenda for state policy will be discussed that suggests a new way of doing business. One of our first tasks is to strengthen the connection between policy-makers and practitioners and to develop much more collaboration between all actors in the system. Every action we take should support student learning.
- Fourth, ECS has entered into a collaborative project with the Coalition of Essential Schools and with five states that is attempting to develop a different policy environment to support what needs to happen in schools.

HISTORY OF RECENT EDUCATIONAL REFORM

Recent literature critical of the American high school paints a picture of schools in which only the very talented or the very needy receive individual attention and in which low academic expectations growing out of implicit treaties between students and teachers are the norm. Teaching is generally characterized as flat and uninteresting content delivered by unengaged teachers whose working conditions discourage a higher level of involvement in their work.

What policies have led to these kinds of schools?

Reforms of the 1970s are most clearly marked by an increase in accountability for students' attainment of basic skills. Reforms included competence-based education, test-based instructional management, and minimal-competence testing (MCT). Over 30 states developed minimal-competence tests—a movement that in hindsight limited student skills attainment. These MCT tests set the educational objectives for the teacher, and teachers began to emphasize the content that they knew would appear on the test. They began to teach in a format that would prepare students to deal with the content as it would be tested. Some teachers even taught items that would likely be on the test. Meanwhile, the rest of the curriculum was de-emphasized. Although most students could perform well on the tests measuring basic skills, these tests were placing a cap on the level of skills that students were learning.

Reforms during the mid-1980s sought to raise educational standards that the minimal-competence movement had depressed. Educational reform has been spurred by major economic development strategies that have viewed a well-educated workforce as a crucial component of economic growth and competitiveness. State-level policy-making became more prominent as states began to provide the majority of funds for education, as policy-makers lost confidence in school districts' ability to pay sufficient attention to curriculum quality and academic standards, and as the federal government withdrew its support for education. This resulted in state reforms that mandated educational outcomes through state assessments, new curriculum frameworks, increased high-school graduation requirements, and teacher evaluation criteria. These reforms have been aimed at the heart of the educational enterprise. What few people realize is the impact of state policies on what is taught and tested in this country: 47 states have some type of curriculum guide in science or mathematics, 49 states have a state assessment program, and 22 states require textbooks to be selected from a state-approved adoption list.

IMPACT OF RECENT REFORMS

In general, recent reforms have had positive impacts on curriculum and instruction in many school districts. For districts with limited resources, state curriculum frameworks have provided guidance for either implementing science curricula where none existed or revising existing curricula in the light of current trends in science education.

Two years ago, with funding from the National Science Foundation, ECS researchers (Armstrong et al., 1988) performed case studies to determine the impact of state policies on science education in schools and districts. ECS visited three states—California, Michigan, and Virginia—which were selected because they had a different mix of initiatives, as well as diverse orientation to state vs. local control. Researchers looked at the impacts of state curriculum frameworks, state assessment, and textbook adoption policies. Case studies were completed in four districts in each state and three schools in each district. Three state policies were studied in depth:

- *Curriculum frameworks.* In science, K-8 frameworks spell out major categories of knowledge, concepts, and processes that students should learn. Most frameworks cover science, technology and society, ethical issues, the nature of science, attitudes toward science, and the major content areas of science—biological science, physical science, and earth and space science. Although there is some diversity in what is included, a major difference among state frameworks is their level of specificity and coherence. Frameworks that spelled out a philosophy and gave teachers model curriculum ideas and strategies had a greater chance of implementation. Frameworks that were seen as of high quality and legitimacy also had a greater chance of implementation.
- *State assessments.* Although most state curriculum frameworks are not mandated, participation in the state assessment program is. The major

incentive for districts to use the state curriculum frameworks is the public reporting of test scores from the state assessment, which is tightly aligned with the curriculum. The design of the assessment affects its impact in classrooms. Matrix sampling of items allows more content to be covered and less opportunity to teach to the test. Reporting of scores within socioeconomic-status bands or comparing results with expected results based on ability scores diminishes the misuse of test scores.

- *Textbook selection criteria.* Twenty-two states, mostly in the South, require districts to select books from a state adoption list. Selection is based on a number of state regulations or criteria, including such items as how well a text covers the state's curriculum framework, copyright date, readability level, and other generic criteria that show up on a long checklist. Although well intentioned, this process does not result in the selection or development of instructionally coherent textbooks. The best analysis of this can be found in Harriet Tyson-Bernstein's book, *America's Textbook Fiasco: A Conspiracy of Good Intentions* (1988).

The use and articulation of these policies resulted in the following impacts in school districts:

- Increased emphasis on science.
- Revision of local curricula and increased teacher training using Title II funds.
- Improvement in the quality of instructional materials.
- Strengthening of ties between districts and schools.
- Increase in hands-on instruction.
- Encouragement of districts to emphasize science, without the result of "teaching to the test."

In school districts studied, ECS found that certain local conditions facilitated the implementation of these policies. These were:

- A district's or school's strong desire to do well on the state assessment, because of "accountability pressure."
- District leadership and a commitment to teaching science.
- A match between the state framework and district philosophy (the district had to "buy in" to the state framework).
- Centralization of science curriculum and instructional materials at the district level.
- Teacher involvement in developing and revising curricula to support the state framework.
- Availability of textbooks, hands-on materials, and activity kits.
- Training of and assistance for teachers.
- Monitoring of implementation by district and school leaders.

Although we found a number of positive impacts of state policies on school districts, a number of barriers remain. Most of these barriers are traditions from a model of education we are using that dates back to the late 1800s. Although state policy has focused on raising the standards of education, policies have only recently begun to be directed toward changing what actually happens in the classroom. Some of these remaining barriers are the following:

- A model of pedagogy that emphasizes lecture, recitation, and seatwork.
- Seven 45- to 50-minute periods per day.
- Teacher isolation.
- Lack of curriculum integration.
- Student tracking.
- An average teaching load of five courses and 150 students.
- The pressure to "cover" an increasing amount of subject-matter content.
- A high-school curriculum designed for college preparation.
- A focus on basic skills, not on ideas or critical thinking.
- Pencil-and-paper testing of content knowledge.
- Limited in-service training for teachers.
- Lack of discretionary resources for classroom teachers.

Removal of these barriers is going to take a new way of thinking about school reform, focusing on systemwide change. Before thinking about new policy tools for systemwide change, it is important to understand the nature of the policy process.

THE NATURE OF THE POLICY PROCESS

Elmore and McLaughlin (1988) provide an excellent description of the educational policy process in a monograph entitled *Steady Work: Policy, Practice, and the Reform of American Education*. They suggest that "educational reform operates on three loosely connected levels: policy, administration and practice. Each level has its own rewards and incentives, its own special set of problems and its own view of how the educational system works" (p. v). Since each is a major actor and translator of policy, the key to the next reform agenda will be to strengthen the connections and dialogue among them and recognize their interdependence.

Elmore and McLaughlin note that "policy consists of authoritative decisions on the purposes of education, on the responsibilities of individuals and institutions, on the money required to run the system, and on the rules required to make it operate effectively and fairly" (p. 5).

> Educational practice consists of the fine-grained instructional decisions necessary to teach the content, manage a classroom, diagnose and treat individual learning

> problems and evaluate one's own performance and the performance of one's students. . . . One survives and performs as a classroom teacher by having strong beliefs about the importance of the task, by developing knowledge of content and process, by developing strong interpersonal skills and by learning how to maintain one's position in an organization. . . . Successful performance, from the teacher's point of view, is likely to be defined in very particular ways: covering a certain amount of material, maintaining order in the classroom, bringing the class as a whole to a certain level of mastery and getting a specific student over a specific hurdle [pp. 5-6].
>
> [Administrators are distinguished from policy makers and teachers by] their preoccupation with the maintenance and development of the organization. One survives and performs as an educational administrator by learning how to juggle the competing demands of politics, organization and practice. . . . It should come as no surprise, then, that administrators tend to identify successful performance with the health of their piece of the system, rather than with the performance of individual students or the performance of schools [p. 7].
>
> The use of policy as an instrument of reform exacerbates differences among policy makers, administrators and practitioners. For an elected official, reform means identifying the problems the public has with education, distilling them into a politically feasible set of remedies and constructing the coalition necessary to turn remedies into policies. [These] remedies . . . must be appropriate for . . . the whole state. . . . Administrators make decisions that extend general policies to particular settings . . . with particular people doing particular things. . . . For teachers, reform means changing established patterns of practice, translating broad and often unclear administrative directives into concrete decisions about how to use time and what to do with this or that student. Teachers understandably see reform policies, . . . from the point of view of getting through the material, adjusting their routines to new supervisors and new roles, meeting new reporting requirements, implementing new testing procedures and communicating new expectations to students. . . . Teachers are often the last to be heard from on the effects of reform policies and the first to be criticized when reforms fail [p. 8].

One reason for the unresponsiveness that policy-makers observe in teachers and administrators is that they have different roles and incentives. Another is that mandates that require broad-scale changes take time to implement. It takes time and adaptive behavior to work out the consequences of reform. Each level of the system must depend on others for knowledge and skill that it does not have.

THE NEW REFORM AGENDA

The new reform agenda builds on what we have learned about people's roles in the system, policy implementation, and a new vision of system-wide change. Lessons from past policy implementation suggest several criteria and strategies for the new reform agenda:

- Larger constituencies for education must be created. Everyone has a stake in the success of our schools, not just parents and educators. Business and community leaders need to be involved in the development of the new reform agenda.

- The interdependence among policy-makers, administrators, and practitioners suggests that reforms will not have large-scale or long-term effects, unless they involve substantial dialogue among the actors in the educational system.
- Reform policies cannot lead to fundamental change in the classroom without addressing changes throughout the educational system. Policies and practices must be rethought from the bottom up and be geared to the most important educational goal: improving student learning.
- Educational reform needs to be grounded in an understanding of how teachers learn to teach, how school organization affects practice, and how these factors affect children's performance. Increasing student learning should be the goal of all reforms.
- Reforms must be based on a strategy that promotes the development of an infrastructure that will support and sustain the reform effort. Anderson and Cox (1988) suggest that a strategy is needed that connects people within their organization, as well as connecting different organizations, to solve educational problems. Actions of groups representing schools, communities, business, higher education, and policy-makers should lead to strategies based on collaboration, not competition; distributed leadership, not authoritarian leadership; flexibility of processes and structures, not rigidity and competition; and approaches to change suitable for fast-changing and complex environments.
- State education agencies need to take on new roles to support reforms in schools and school districts. New roles include helping schools and districts to build greater capacities; developing new forms of assistance that directly help teachers, schools, and school districts; requiring less monitoring and regulation while providing greater measures of school and system accountability; and providing financial incentives to encourage systemwide change.
- Experimentation and risk-taking in schools need to be supported through incentive or planning grants to school districts and schools and through waivers from existing regulations that get in the way of innovation.
- Decisions should be moved closer to the classroom. Practitioners should be charged with the development of solutions, rather than having requirements mandated.
- Decentralizing school districts and moving toward site-based management and shared decision-making should lead to an educational system that better meets the needs of increasingly diverse students.
- Enough time and support for implementation of new practices must be allowed. Behavioral change and change of practice take time. System change does not lend itself to "quick fixes."
- Finally, there needs to be an opportunity for reflection. Individuals

need to step back from the daily routine to consider the larger purpose of their actions, the connections and fit between their actions and those of others, and next steps.

These are just some of the new ways of doing business to begin a restructuring of the educational system. Below is a description of an ECS project that is attempting to bring many of these ideas together in an effort to rethink the educational system in five states.

DEVELOPING NEW POLICY ENVIRONMENTS

A new project at ECS, Re:Learning, is a collaboration among ECS, the Coalition of Essential Schools at Brown University, and schools and policy-makers in five states. This project is a long-term commitment of the partners and is designed to take advantage of the best information available from reform efforts about how students learn and what system changes are needed to develop and promote leadership, creativity, and thoughtfulness among students and educators alike. The work of Re:Learning has three components—in the schools, at the state and district level, and nationally.

Schools involved in Re:Learning are committed to the nine principles that unite Ted Sizer's Coalition of Essential Schools. Good schools are never exactly alike, but they do have these common characteristics:

- *Intellectual focus.* The school should focus on helping students to learn to use their minds well. It should not attempt to be "comprehensive" at the expense of the school's central intellectual purpose.
- *Simple goals.* The schools' goals should be simple: that each student master a limited number of essential skills and areas of knowledge.
- *Universal goals.* The schools' goals should apply to all students, although the means to the goals will vary as those students themselves vary. School practices should be tailored to meet the needs of each group.
- *Personalization.* Teaching and learning should be personalized to the greatest extent feasible. To that end, a goal of a maximum of 80 students per teacher should be vigorously pursued. Decisions about curriculum, allocation of time, and choice of teaching materials and their presentation should rest unreservedly with the school's principal and staff.
- *Student-as-worker.* The governing metaphor of the school should be student-as-worker, rather than the more traditional teacher-as-deliverer-of-instructional-services. A prominent pedagogy should be coaching, to provoke students to learn how to teach themselves.
- *Diploma by "exhibition."* The diploma should be awarded on successful final demonstration of mastery—an "exhibition"—of the central skills and knowledge of the school's program. The familiar progression

through strict age-related grades and "credits earned" by "time spent" in class will be unnecessary.

- *Attitude.* The tone of the school should explicitly and self-consciously stress values of unanxious expectation ("I won't threaten you, but I expect much of you"), of trust (until abused), and of decency (fairness, generosity, and tolerance).
- *Staff.* The principal and teachers should see themselves as generalists first (teachers and scholars in general education) and specialists second (experts in one particular discipline).
- *Budget.* Ultimate administrative and budget targets should include a maximum of 80 students per teacher, substantial time for collective staff planning, competitive staff salaries, and a per-pupil cost no more than 10% above that of traditional schools. Inevitably, this will require the phased reduction of some services now provided in many comprehensive high schools.

After a school faculty, with the support of the school and district leadership, decides to join a state's Re:Learning effort, it develops a comprehensive plan for reshaping the curriculum and the school and prepares a timetable for implementation. The design of each school—the actual detailed functioning of its program—while consistent with the Coalition principles, will evolve to suit the community, the students, and the teachers. State, district, and private funds will support staff planning and school-initiated staff development activities as the staff implements its plan and gradually redesigns the school to be consistent with the Coalition's principles.

Schools will document their activities to allow assessment of their work and to serve as a source of data for other schools to use. Schools also agree to meet jointly with other Re:Learning schools and with Coalition schools nationally to share information, observations, and techniques. On joining the Re:Learning initiative, schools accept an obligation to support work among partner schools.

An in-state school coordinator will assist each participating school with restructuring and serve as a continuing "critical friend." The coordinator will also organize workshops, bring teachers from different schools with common concerns together, troubleshoot with district and state officials, and represent the group to the public and to the media.

Coordinators will most often be located at universities to help to build the link to higher education and bring Re:Learning concepts into the education of teachers and administrators. The coordinators will receive intensive training and continual consultation from the Coalition and ECS and will maintain contact with Coalition central staff. They will also work closely with state education-agency personnel.

State and district leaders involved in Re:Learning will focus their attention on making changes in administration and policy that reflect a commitment to a truly "bottom-up" system—i.e., one in which the driving force for actions of people, from teacher to governor, is helping all students to learn to use their minds well.

Just as schools have principles to guide them throughout the Re:Learning effort, so too do administrators and policy-makers have a set of principles that they will follow:

- *Build a new vision of education.* The public, business and state leaders, and education professionals should build a new shared vision of an educational system where all students have an equal opportunity to use their minds well and where there is a commitment to using the best available knowledge to provide meaningful teaching and learning experiences.
- *Organize on behalf of student learning.* The roles and responsibilities of adults and the allocation of resources should be productively redesigned primarily to support the best learning for all students, not bureaucratic or political interests.
- *Create new working relationships.* Collaboration, shared leadership, and mutual responsibility should serve as the dominant basis for establishing working relationships throughout the educational system.
- *Develop a culture of learning.* Adults throughout the system should see themselves as continual learners and problem-solvers, rather than purveyors of "right" answers and standardized solutions.
- *Act with regard for people.* Long-term and short-term actions to rebuild the educational system should be balanced in ways that treat people with dignity and respect.

Links between the two sets of principles bring a coherent focus to the activities of teachers, administrators, and policy-makers at all points in the system. Decision-making and incentives at all levels should encourage students and adults alike to use their minds well and to engage in meaningful activities. ECS will assist in the establishment of a state-local cadre—a group of key opinion leaders drawn from school level, district and state education staffs, universities, and other appropriate decision-making groups—to focus on changes that need to be made in state and district operations and in universities and colleges to allow the school redesign to flourish. Particular attention will be directed toward issues of student assessment, accountability, leadership, and resource allocations. The cadre will generate and help to guide systemwide changes and strategies.

On the state policy level, ECS will assist the steering committee (which is expected to include the governor and chief state school officer, the state higher-education executive officer, legislative leaders, university leaders, the state board chair, and other state and local leaders) as it considers how to develop public support for school redesign and how the policy-making environment needs to change to make redesign of the full educational system possible in its state.

As part of their responsibility in Re:Learning, state and district education leaders agree to devote sufficient resources to participating schools to allow teachers planning time, allowances for travel to other schools, and other assistance. A typical amount would be about $50,000 per year per school to initiate the effort. The funds could come from a combination of state, district, and private sources and may be new dollars or reallocated funds.

As the Re:Learning initiative is put in place, a documentation and analysis design will accompany it to interpret the new structure and assess progress appropriately, to help to guide future action and thinking.

The national component of Re:Learning will support the state-by-state and school-by-school changes. A major force in moving state action forward is creating a national debate and discussion on changes needed in the educational system so that the "best of learning for all students" is the driving force for change. Cross-state groups, nationally recognized leaders, and the mass media need to be formulating, presenting, and debating an effective message that changes the language of what education can and must be.

SUMMARY

Barriers to high-quality biology education will be removed as barriers to changing the educational system are removed. Efforts to improve education must be based on collaborative efforts to develop a new vision, a supporting policy environment to promote effective actions, and opportunities for reflection on actions taken. This will be accomplished when all actors in the educational system base their actions on the educational bottom line—improving student learning.

REFERENCES

Anderson, B. L., and P. L. Cox. 1988. Configuring the Education System for a Shared Future: Collaborative Vision, Action, Reflection. Denver: Education Commission of the States; Andover, Mass.: Regional Educational Laboratory for Educational Improvement of the Northeast and Islands.

Applebee, A. N., J. A. Langer, and I. V. S. Mullis. 1989. Crossroads in American Education: A Summary of Findings. Princeton, N.J.: Educational Testing Service.

Armstrong, J., A. Davis, A. Odden, and J. Gallagher. 1988. The Impact of State Policies on Improving Science Curriculum. Denver: Education Commission of the States.

Elmore, R. F., and M. W. McLaughlin. 1988. Steady Work: Policy, Practice, and the Reform of American Education. R-3574-NIE/RC. Santa Monica, Calif.: RAND Corporation.

Tyson-Bernstein, H. 1988. A Conspiracy of Good Intentions: America's Textbook Fiasco. Washington, D.C.: Council for Basic Education.

30

Different Schools: Same Barriers

GRACE S. TAYLOR

Before coming to the Education Department at Brown University last year, I taught biology for 18 years at a comprehensive, urban high school of 2,600 students. This is a personal reflection about a profession I love. I borrow the documentation for my remarks from my experience and the experiences and insights of colleagues around the country.

High schools throughout the United States are remarkably uniform in organization and in their approach to teaching biology. Here and there, some creative administrators and gifted teachers manage to offer a real science experience to their students, but for the most part, biology is rote learning and cookbook laboratory experiments. Thus, the window of opportunity to help students to make sense of the living world is closed. In this discussion, I am concerned with the question: What is it that inhibits change in the biology classroom?

SCHOOL TRADITION

It is because of tradition that change in high schools is slow and difficult. As Ted Sizer traveled around the country visiting schools, he particularly

Grace S. Taylor is a clinical professor of biology (education) at Brown University. She received a B.A. in biology from Emmanuel College and an M.A. in biology from Wellesley College. In 1969-1987, she was a biology teacher at Cambridge Rindge and Latin School, in Massachusetts. She was chosen an outstanding biology teacher in 1984 by the National Association of Biology Teachers.

focused on observing biology and history classrooms. He comments, "It got so when I visited biology classes, no matter what school it was, I could almost predict what would be taught depending on what month it was. It was so similar. They were following the textbook" (T. Sizer, Brown University, personal communication, October 3, 1988).

Structure of the Day

In many schools, classes are run in the same time block year after year. In a seven-period day, teachers are scheduled for five classes, each of which meets at the same time each day, usually for 45 minutes. Only the honors and upper-level sciences may have a weekly double period for laboratory activities. For maximal efficiency, each classroom is scheduled for use each period, so the teacher who is "free" for a particular period must go elsewhere. The classroom is not his or her own, nor is it available for a student to finish a project. For the biology teacher, this necessitates setting up laboratory experiments before or after school. But in many city schools, teachers are not allowed to stay after school, for safety reasons. All these factors lessen the ability of the teacher to give quality instruction.

Teacher Schedule

In most school districts, the maximal number of students a teacher may have in a class is 30; therefore, he or she may be responsible for a total of 150 students. Some school systems realize that this is untenable, particularly in a laboratory setting, and have lowered the number to 25 or even 20 per class. Experts in education know that in order for students to learn most effectively they must be actively engaged. Authentic inquiry laboratory exercises where the students are involved in the process of experimentation and investigation often cannot be implemented within the rigid timeframes of the day and the teacher's schedule.

Think of a single teacher conducting a pond study with 25 students at 25 microscopes (if she is lucky enough to have a microscope for each student). Of course, she wants them to make their own wet-mount slides from the water she and the class have brought in. She wants them to have the thrill of seeing live paramecia and vorticellae. But there will be constructive chaos in the classroom, and 3 minutes after the bell rings another teacher's class will be coming into the room. In thinking about this lesson, the teacher must make a choice between original discovery and static learning from prepared slides. The latter will fit into the confines of the day easier.

Teacher Isolation

The structure of most high schools means that teachers do their work behind closed doors. Exciting ideas and methods may be born, but generally stay behind those doors. The bell rings, the teacher does corridor duty, and then the next class comes in. There is rarely any attempt to organize the schedule so that teachers who teach the same subject can have common planning time. Even in schools where there are department examinations, teachers usually do not discuss together what is essential for their students to know. Instead, someone may make up the examination this term, or each teacher may send in questions on a specific area. The synergistic effect of many stimulated teacher minds working together is lost. No common goals of the biology curriculum are debated, no consensus reached.

There are actually subtle institutional disincentives to collaboration. If one spends a period talking to a colleague about what is happening in class, one is taking time away from preparation, planning, and correcting for the next class, the next day. Yet it is necessary to have horizontal (departmental, 9-12) and vertical (districtwide, K-12) articulation in order for curriculum change to be discussed and eventually implemented.

Teacher Instruction

Most high-school students find biology "boring, because you have to memorize too much stuff." The curriculum is so overwhelming that it seems to dictate the pedagogy. Faced with feeling that he or she has to deliver too much content in as efficient a way as possible—a 700-page text in 180 schooldays—the teacher resorts to lecturing. Depth is subjugated for breadth, and coverage is confused with real study. Large conceptual themes, such as relationships among living things, tend to be ignored in the pursuit of specific structural terminology.

And who are the students to whom this barrage of information is directed? College-bound students eager to learn biology in detail? It may be surprising to hear that there is a good chance that more students who took freshman science will, 8 years later, have spent time behind bars than will have a bachelor's degree in either biology, chemistry, physics, earth science, or science education (Leyden, 1984). Teachers must know who their audience is and teach appropriately.

The tradition has been the teacher talking, not the student doing the work. The kind of instruction that is needed is one that engages students' minds, builds their skills, and helps them with problem-solving strategies. This may require both restructuring the schoolday and viewing the teacher in a new way—as a helper, not a dispenser.

Artificial Separation of the Sciences

There is too much specialization in the high-school science curriculum. The typical high-school sequence is earth or physical science, biology, chemistry, and then physics. It is as though each science teacher is preparing the students for a career in one of the sciences, instead of helping them to gain the knowledge and skills that they will need in order to understand science in our society. In biology, we teach the parts of the eye; in chemistry, the binding of molecules, such as the visual proteins; and in physics, the action of light waves. Because we teach every aspect separately, students keep it separate in their minds, and thus necessary connections are not made and real understanding is lost. An interdisciplinary curriculum would reduce the "I can't do science" refrain that is often heard in high schools.

State Certification

Certification requirements encourage future teachers to specialize in only one of the sciences and thus perpetuate the separation. Most biology majors do take college courses in chemistry, physics, and mathematics and thus could, with the help of some innovative materials, teach in an interdisciplinary way. Some, of course, do just that. (Note that the English language-arts curriculum is composed of five elements: oral language, active listening, composition, grammar, and literature. Each has a separate body of skills and some would say a separate scope and sequence; yet English language-arts teachers are expected to, and do, teach all elements in each of their courses.) The sciences overlap as well. Can one really separate biology and chemistry? Should change involve science teachers as generalists? Interestingly enough, the one science certification that is interdisciplinary is general science, the course that is lowest in the hierarchical scale and is usually taken by the non-college-bound.

Sex Roles

Sixty-three percent of biology teachers are male (Champagne and Hornig, 1987, p. 215). The blatant "I don't hire women science teachers," as one principal said to me in 1969, has been replaced with more subtle forms of discrimination. The only female teacher on the science staff in a high school outside Boston is "locked out of the stockroom." The first female science teacher hired in a high school in New Hampshire (in 1986) left after a year of isolation and was replaced by two men. There are few examples of women scientists in the texts or in the curriculum. If the ideas of women teachers are not held in high regard by their male colleagues, any dialogue involving change will be difficult.

Tracking

Homogeneous grouping makes assumptions that are not true. It assumes that students cannot all learn the same curriculum. It assumes that in a heterogeneous classroom one has to teach to the middle, and the top group is bored and the bottom one is lost. It assumes that it makes it easier for the teacher to teach appropriately, whereas, in reality, he or she has three levels of biology in three textbooks, with three preparations every night and, possibly, three laboratories the next day.

But as long as teachers believe, and thus the students do too, that only certain students can really learn and do biology, there will be no impetus for change. College-bound students know what they have to do, and they will do it, even if they are bored. The advanced-placement level of performance is very specific and is geared to the outcome desired. If students in a lower track are lucky, they may have a teacher who puts emphasis on relevance, process, and skills; but more often than not, the less-able teachers get the supposedly less-able students, and more trees are sacrificed to satisfy the dominant method—worksheets!

TEXTBOOKS

For many teachers, the textbook *is* the syllabus, with a resulting overreliance on the textbook in most biology classrooms. Those of us who have read the literature over the years are aware that the vast majority of teachers of biology and other subjects feel that most of their teaching problems would be solved if only they could find the "right" textbook. With the exception of the Biological Sciences Curriculum Study (BSCS) series (whose latest editions look more and more like the traditional ones, with fewer and fewer open-inquiry laboratory experiments), most of the texts are interchangeable.

It is naive to assume that, should a new and innovative biology approach be proposed, publishers would be interested. Witness the fact that a new social-studies curriculum, *Man: A Course of Study* (MACOS), was rejected by publisher after publisher in 1967. "They told the MACOS developers that their stress on inductive methods, small-group instruction, the teacher as participant rather than authority, and multimedia design were formidable obstacles to adoption by teachers" (Lazerson et al., 1985, p. 36). We cannot assume that things have changed in schools or in the publishing industry.

The demands of state textbook-approval boards also limit teachers in selecting texts that they feel are appropriate for their students and consistent with their goals.

MONETARY CONSTRAINTS

Whenever change is suggested, the "what will it cost?" question is usually the first one asked, and the answer to it alone determines whether the change will be implemented.

Resources

A set of textbooks for a class of 25 students may cost over $400. There is the additional cost of laboratory books, supplies, and perhaps new equipment. The cost of implementing a biology course of study is considerable, and other departments in the school watch the proportional allocations of funds closely.

Time

The old adage, "time is money," is applicable in education as well. In order to look at the curriculum or pedagogy in new ways, it will be necessary to offer teacher training during the schoolday with class coverage by a substitute. This may need to be a continuing process. A colleague gave a 1-day in-service session to biology teachers in his city during which he introduced an original 2-week unit that he had written, exploring land use and decision-making. The teachers were interested in the new approach, but monetary constraints were such that there was no time available for them to practice together and work through the unit. Thus, no real implementation could be expected, nor was it achieved. To the school district, the cost is not merely that of the substitute, but also the per diem of the teacher who is away from the classroom.

Average Teacher Load

The average teacher load is carefully calculated by assistant superintendents in charge of finances to determine the cost-effectiveness of the teacher-to-student ratio. Teachers are expected to teach a full load, at least 100 students and 25 class periods per week. My class schedule was 22 periods a week, because my advanced-placement class met seven periods a week. The principal informed me that I would have to take study hall three times a week to bring me up to the required load. My arguments included the time needed to correct the many essays that are part of the course, the time needed to set up laboratory experiments, and the philosophy that advanced placement should be considered as a double course. They were in vain, and I was assigned to a study hall in the cafeteria three times a week. The irony is that my advanced-placement course and the results achieved in it were often used by the school in public-relations forums.

Classes with fewer than 12 or 15 students are not considered to be cost-effective and are dropped from the schedule. A graduate student in English in our teacher-education program took a field biology course during her senior year in high school. She was very excited by her first taste of real science and said, "Although my project wasn't that original, I really learned a lot. But they didn't offer it again the next year, because not enough students signed up." (It will be interesting to see whether this course, which involved authentic inquiry, will be reinstated, now that its teacher is the superintendent of schools in this district.)

Conferences

In many school systems, not only are teachers not reimbursed for conferences attended, but they are not allowed to go during schooltime. This short-sightedness remains, even though many teachers affirm that conferences are exciting, stimulate creativity, and are more directly relevant to their teaching than further coursework. National Association of Biology Teachers meetings include workshops that are led by teachers and are therefore rich in practical ideas. Yet districts still require only coursework for incremental wage increases. Often, it is not even necessary that courses taken after a teacher has been granted tenure be related to the teacher's subject area.

STANDARDIZED TESTING

Another factor that drives what is taught in biology classrooms around the country is state tests, College Board tests, and high-school departmental tests. The rationale for each may be different, but the end result is conformity and a lack of "open-endedness." In Massachusetts, science tests were given to students in fifth, eighth, and eleventh grades, and the scores were later published in the *Boston Globe*. The results caused some measure of consternation among the science staff in my former high school. They asked: What sorts of things were on the test? What did they have to teach so that the students would do better on the test? It is a classic case of the tail wagging the dog—standardized tests determining the curriculum!

Duckworth (1984, p. 19) stated, "What is dreadfully missing from a standardized test of biology, say, is any real conception of what the study of biology is: there is no way to tell whether a student has a sense of the questions that biologists ask, how to go about exploring them, how they relate to each other, how mistaken hypotheses can be productive."

UNIVERSITY PREPARATION

There is often more than a grain of truth in axioms, and "teachers teach as they were taught" is truer than most. In college classrooms throughout the country, Biology 101 students are trapped in a maze of facts and a haze of terms. They become passive learners, recipients of information, whose habits of thought and inquiry are underdeveloped. Thus, we should not be surprised when our biology student-teachers teach in the same way. Where have they been exposed to dynamic teaching? Where have they learned to pose authentic questions? This summer, as my candidates for the master's degree in teaching were getting ready to do their first teaching at Brown Summer High School, it was clear that they thought of their role as dispensers of information. Lecture, laboratory, and oral question and answer were the only teaching techniques with which they were familiar in the science classroom.

PRESERVICE EDUCATION

If the teaching in university biology classes has been less than inspired, it behooves teacher-education programs to develop teacher methods that are student-centered, are interactive, and serve as catalysts for student thinking and problem-solving. Unfortunately, teacher candidates are often taught by professors who are decades away from actual classroom teaching and who lack first-hand knowledge of the current school populace. There is a substantial incongruence between the way we are teaching science today, with our emphasis on reading a polysyllabic textbook and answering laboratory workbooks with packaged questions, and the learning styles of the students we teach. Is there too much emphasis on the well-written lesson plan and maintaining classroom control, and too little emphasis on developing methods to help students to use their minds well?

Teachers often take for granted that their students are like them in the way their minds work, in the way they think and feel. "When teachers are working with students *who are very much like themselves*, there is relatively little to learn about teaching, at least insofar as technique is concerned, that is not supplied either by common sense or by the knowledge of the material to be taught. But when teachers and students are *not* alike in important ways . . . the knowledge called for under those circumstances is genuinely knowledge about teaching *per se*" (Jackson, 1986, p. 26). Preservice work with a multicultural component is imperative.

There is much that educational research can offer both the new and the experienced teacher, but the researcher and the teacher often do not speak the same language. As a colleague so aptly put it, "There is a need for translators."

PROFESSIONAL IN-SERVICE EDUCATION

Too few school systems set aside release time for teacher collaboration, updating, training, or visiting other classrooms and schools. After teachers have supervised homeroom and taught approximately 125 students, it is too much to expect that an after-school meeting to discuss change in the biology curriculum will be productive, let along innovative. Even when release time is available, teachers' attitudes may be negative, as past experience has shown them that they will not be involved in setting the agenda. What will transpire will not be relevant or meet their needs.

When teachers are fortunate enough to go to an all-day workshop, such as the ones held by the Institute of Secondary Education at Brown, they say, "I come back to the classroom feeling invigorated." The students can only benefit from such a recharged teacher, but new content must be linked with new teaching techniques. As a colleague said, "We must not only use that knowledge, but make it sing!"

There is no doubt that the success of the BSCS curriculum was aided by workshops sponsored by the National Science Foundation. If such an initiative is begun again, those who plan the scheduling must take into consideration the varied roles of women, so that as many women teachers as possible will be able to participate.

ADMINISTRATION

Leadership in the central office and in the high school is often myopic. In a suburban high school, the science-department head is serious about deleting double laboratory periods when he says that "smart kids don't like to do labs, because they know what the results will be and, besides, it makes scheduling difficult." It makes one wonder whether he thought of changing the *types* of laboratory experiments the students did, rather than the structure of the laboratory itself. In my former school, science teachers are now required to babysit a homeroom, instead of having that time free to set up for laboratory class, as they did before. Teachers yearn for creative curriculum directors who have a knowledge of the discipline, its trends, and the teaching strategies needed to encourage the students to true learning. They want a leader with vision and the ability to excite the staff to work toward shared goals. Alas, most are pedantic and have little knowledge of good science education and so exert little or no direction. If any change is going to occur, it may have to be teacher-initiated or teacher-directed.

But teachers are swamped with "administrivia"—homeroom, late slips, absence lists, cut slips, inventory forms, student grading-policy agreements, substitute folders, study halls, etc. Organizing for a guest speaker or a field trip can become so difficult (with forms needed to be signed by three different administrators) that many teachers just do not bother. Their energies have been directed toward the wrong areas.

TEACHER ATTITUDES

The teacher is the crucial link in any conversation about change, but change is both a challenge and a threat. Are a majority of biology teachers convinced that there is a need for change, or are they in an "if it ain't broke, don't fix it" mentality? Many feel overwhelmed by the explosion of biological knowledge in the last 2 decades, as they try to keep up with recent advances. They feel confused by the reports that suggest that they must teach more and the data that show that American students understand less science (Education Week, September 28, 1988).

Does the aging teacher population have a desire for renewal? From where will the impetus for change come? In the 1960s, the reformers came from the country's prestigious universities, and they modeled the curriculum after the academic content of college courses, with little concern about the realities of classroom teaching (Lazerson et al., 1985). If, indeed, change is in the air again, will it come from presentations at conferences like this, where there appear to be only two active teachers on the many panels? Teachers rightly rebel at the imposition of change, particularly from those they feel are not cognizant of the realities of the high-school classroom. "New math" was born, and it died. That could well happen to a new biology curriculum that is initiated "top-down."

There are teachers who feel that problems in the biology classroom are not the result of the curriculum itself, but occur because students are unmotivated and "functionally illiterate." This seems to be a classic "blame the victim" ploy. Perhaps one could ask whether the teachers are "methodologically illiterate."

Historically, when curriculum change has occurred, there has been insufficient teacher training. It would seem imperative that new curriculum be linked with new pedagogy; but while teachers are open to learning new facets of their subject, many consider professional development degrading: "I already know how to teach!"

There are many reflective teachers who are uncertain about what is best for their students—how the students learn, what can be done to motivate them, how to develop the skills they need. A 17-year biology veteran with whom I work stated, "I'd realized the past few years that the kids had changed, and what I used to do in class no longer worked, but I don't know what else to do." It is the openness to change that must be nurtured and developed.

CONCLUSION

Reflecting on all these barriers to change has been, by necessity, a negative exercise. Experiences and anecdotes from many schools validate the barriers to change. In thinking about changing the curriculum, we must ask not only what it is that we want our students to know, but also how it should best be taught. Change necessitates restructuring the day and the teachers' schedules, modifying certification requirements, developing creative administrators, improving curriculum models, rethinking teacher education, and building in time so that teachers can have a true dialogue. If change in the curriculum—and, by necessity, in the pedagogy—is deemed necessary, then the crucial question to ask is: "What are the incentives for districts, administrators, and teachers to change the way the biology curriculum is approached?" Get the incentives right, and the rest will fall into place!

ACKNOWLEDGMENTS

I owe special thanks to Ann Beachan, Jonathan Bealer, Paula Evans, Amy Gerstein, Judy Johnson, Gordon Mendenhall, Theodore Sizer, Nancy Topalian, Sharon Wolff, and Carolyn Wyatt.

REFERENCES

Champagne, A. B., and L. E. Hornig, Eds. 1987. The Science Curriculum. Washington, D.C.: American Association for the Advancement of Science.
Duckworth, E. 1984. ". . . what teachers know: the best knowledge base . . . ." Harvard Educ. Rev. 54:15-20.
Education Week. September 28, 1988.
Jackson, P. W. 1986. The Practice of Teaching. New York: Teachers College Press.
Lazerson, M., J. B. McLaughlin, B. McPherson, and K. Bailey. 1985. An Education of Value: The Purposes and Practices of Schools. New York: Cambridge University Press.
Leyden, M. B. 1984. You graduate more criminals than scientists. Sci. Teach. 51:27-30.

PART VII

Accomplishing Curricular Changes—Implementation

31
Problems and Issues in Science-Curriculum Reform and Implementation

PAUL DEHART HURD

There is nothing more difficult to manage, more dubious to accomplish, or more dangerous to execute than the introduction of a new order of things.

[Machiavelli, 1977 (1513)].

This nation is once again demanding a reform of education with attention directed especially at deficiencies in the teaching of science and mathematics (Hurd, 1984, 1985). The charge implies that young people are being ill prepared for living in an "information age" and for meeting the social and economic demands of the twenty-first century (NAS, 1982; NSB, 1983; National Commission on Excellence in Education, 1983). In the last 5 years, 1983-1988, over 100 national commission, panel, or committee reports have been published, in addition to dozens of books by informed educators—all critical of precollege education in the United States. It should be noted, however, that the vast majority of reports were developed by citizen groups, government agencies, economic organizations, or business or industry, and not by schools or educators.

The need for educational reform has been viewed as a national crisis, and immediate action has been demanded. Leadership for the reform was assumed for the most part by politicians, particularly state governors (ECS,

Paul DeHart Hurd is professor emeritus of science education at Stanford University. Dr. Hurd, long a leader in science curriculum developed for the schools, is a member of the human biology program under development at Stanford.

1983; Kirst, 1984), and by business and industrial organizations (CED, 1985). Currently, a number of private foundations are studying critical aspects of the overall school-reform effort, such as urban educational problems and education of teachers (Carnegie, 1986; Ford Foundation, 1984). Changes in the subject matter to be taught and its context are being explored by several science-based groups (the National Science Foundation; AAAS, 1987; ACS, 1988; NASTS, 1987).

The various science teachers organizations have been cautious about entering the debate on curriculum reform. A few of the organizations have used ad hoc committees to refine previous statements of science-teaching goals. These organizations have been active in forming networks, alliances, or coalitions among teachers to share ideas about what should be done to improve the condition of science education, but to what ends is not clear. A 1988 study of articles in 12 leading science-education journals—such as *The Science Teacher*, *The Physics Teacher*, *Journal of Chemical Education*, *Science Education*, and *American Biology Teacher*—in 1983-1988 found that only 22 of 4,884 feature articles were responses to the concerns represented in the national reports on educational reform. Of the 22 articles, 16 stressed the importance of including technology in science courses and four recommended including scientific-societal issues. None of the science-education journals carried an article that systematically explored the scientific and social issues that underlie demands for a reform of science education (Hurd, unpublished data).

The 1980s are not the first time in this century that attempts have been made to redirect the teaching of science. Reform issues arise whenever a perceived economic or social crisis appears on the American scene, such as the shift from an agricultural to an industrial economy or, as is now the case, a shift from a "postindustrial society" to an "information age." Periods after wars always generate concerns about what should be the nature of public education. World War II led to renewed attention on precollege science education with the goal of strengthening the U.S. technical workforce (Steelman, 1947). Some education reform movements are politically inspired, for example, by the successful launching of Sputnik by the USSR in the 1950s (President's Science Advisory Committee, 1959) and by the Japanese domination of the global economy in the 1980s. Politicians take the stance that schools must be doing something wrong, or the United States would be first or on top of the situation. A persistent theme in the 1980s reform movement is that the United States has lost its competitive edge in world markets and therefore should revise the school science curriculum. Schools are called on to initiate a new social contract with the nation—one that redefines standards of excellence and will serve to turn the tide in the country's eroding foreign economic competition. It is frequently suggested in the public press that we should adopt the science curriculum of our chief competitor, Japan. Japan, however, is in the process of reforming its educational system to ensure that it will not lose its competitive position in the world (Hurd, unpublished manuscript).

Bringing about a fundamental change in the science curriculum is a complex process. In fact, it is a process that has yet to be resolved. A major reason for this situation is a tendency in the United States always to deal with problems, rather than first identifying and interpreting the underlying and interacting social, cultural, and scientific developments that project new educational demands.

A brief look at some current efforts to foster educational changes will demonstrate why the movement is failing so far. One action has been to use the public press to deliver the worst bashing that schools have ever had to endure. Teachers are portrayed as incompetent, students as ignorant of whatever you may regard as important, school principals as not knowing how to provide leadership, schools as not being administered in a business-like fashion, and students' scores on standardized tests as an indicator of poor teaching. A common means for dealing with these problems is to reduce financial support until schools do better.

Another policy has been to legislate change. Within the last 5 years, over 800 laws, mandates, or regulations have been established by states to influence practices in schools. On the one hand, requirements for teacher certification are increased for graduates of teacher-education institutions; on the other hand, there are lower qualifications for any citizen who wishes to teach and has had little or no training.

The most common recommendation for educational improvement is for everyone concerned to try harder. This idea is implemented by requiring more of everything: more schoolhours per day and more schooldays per year, more rigorous courses (a euphemism for "rugged"), more testing of both teachers and students, more "time on task" in classes, higher standards for getting into college, more involvement of business and industry and of university faculty in school affairs, more laboratory work in science classes, more use of computers and other electronic technology, more publicity for "good" or effective schools and more "bad" publicity for ineffective schools, more in-service training for teachers and principals, and so on. About the only "less of" recommendation is less opportunity for students to participate in competitive sports or other extracurricular activities if they do not meet certain academic standards. There may be merit in some of these recommendations, but in the aggregate they reinforce the conditions and circumstances that give rise to the quest for educational reform in the first place.

What have been the results from these strategies? Teachers are demoralized, parents disillusioned with schools, and students "turned off" by science; and there is a growing attitude that it is probably better to go back to traditional curricula and modes of instruction and learning. Considerable publicity has been given to "effective schools," schools that appear to be doing something better than they did in the past. I have searched the published reports on these schools, and I did not find changes in their philosophy of science education, a recognition of the impact of modern science and technology on society, or evidence that student learning was more productive.

A reform of high-school biology has been under consideration for nearly a century. At roughly 10-year intervals, a committee is formed with new perspectives on the teaching of biology (Hurd, 1961; Mayer, 1986). Conferences are convened, resolutions passed, reports published, a few workshops given for teachers at regional or national conventions—and soon all are forgotten. A few years later, the cycle is repeated; but there is no review of the accumulated history that might lead to a new conceptual framework for an education in biology. Compare, for example, the report of the Committee of Ten, Twelve, Fifteen (NEA, 1894) with *A Nation at Risk* (National Commission on Excellence in Education, 1983). They are similar in their recommendations. Neither of these reports has as yet stimulated the development of a biology curriculum that recognizes the issues identified by the reformers. And it can be added that none of the other national reports on the improvement of science education published in the 1980s has so far brought about significant change in what is taught in schools.

A good deal of the ineffectiveness of the national reports is inherent in the reports. As one reads these reports, one realizes that they tend to be more critical than creative, more speculative than informed, more slogans than solutions, more visible than valid, and more problem-directed than issue-directed. Their positions on education tend to be supported by passionate rhetoric and uncertain statistics. The educational slogans of "quality," "excellence," and "scientific literacy" have been around for more than a century and are still wanting in definition.

The central problem is how to introduce into schools a biology curriculum that represents the ethos of modern biology, ensures more productive learning by students (Resnick, 1987), considers social changes and cultural shifts, and is in a context that has educational validity for the foreseeable future (Cole and Griffin, 1987). All biology-reform committees over the last 100 years have failed in attempts to implement a curriculum in which the goals were the proper education of a citizen in the sense of being better informed about life and living, more concerned about biosocial problems, and more competent and confident in reaching decisions. This is a much more difficult task than educating scientists and technically trained journeymen to carry out the practice of science.

There is a plethora of reports indicating quantitative deficiencies of science education, but nowhere is there to be found a unifying theory of either science or biology education that has a modicum of consensus (IEA, 1988; Raizen and Jones, 1985; Buccino et al., 1982). Efforts to bring about a reform of science education that proceed "ahistorically" and "aphilosophically" have no anchors in reality and no flag to follow. The most difficult phase of implementing a reform of science education is changing the prevailing beliefs of teachers, parents, school administrators, and school-board members about what an education in science ought to mean. A lack of such a statement of belief only serves to create more confusion than insight and neutralizes reform efforts.

A well-recognized principle in social psychology is that effecting change in an institution requires that all the actors be considered. For schools, this means not only teachers, but parents, students, principals, top administrators, school-board members, politicians, and college and university faculty members in the sciences and in education. In the science-curriculum projects of the 1950s and 1960s, only the scientists and a few token teachers were involved in developing the curriculum rationale and choice of subject matter. All other teachers were to be trained in various types of institute programs taught by scientists who were not involved in producing the materials (Hurd, 1969). School administrators, parents, and students alike were left out of the picture. So were the science educators in colleges and universities, with the result that the next generation of science teachers were never trained to implement the new curricula. The same situation occurred in the departments of science in colleges and universities. These departments are responsible for 85% of the requirements for the certification of a teacher, but they did not pattern course requirements in ways that will improve public education in science. A lesson from the science-curriculum improvement projects of the 1950s and 1960s is that $1 billion for teacher in-service programs and nearly $150 million for new instructional materials will not ensure the success of an intended reform. A study by the U. S. General Accounting Office published in 1984 concluded that the institute programs of the 1950s and 1960s for the retraining of science teachers were largely ineffective (GAO, 1984). Science courses are taught today in the way they were in the 1940s and with the same goals in mind.

Serious blocks in implementing a new curriculum are the misconceptions that teachers have about the various ways of knowing in the sciences and what is meant by knowledge and wisdom. Using biology as an example, when T. H. Huxley, in 1878, developed a biology course for use in high schools, the prevailing theory of learning was known as formal or mental discipline. The underlying assumption was that the mind had a number of distinct and general powers or faculties, such as memory and observation, and that they could be strengthened and developed by mental exercise. Because of the extensive terminology and taxonomy—much of it ideally

Latinized—biology was considered an ideal course for training memory and observation. One needs only to examine a modern textbook in life science or biology to find that the theory of formal discipline still prevails in practice. Most textbooks are little more than beautifully illustrated dictionaries. Note also the number of recommendations in the current science-reform movement that stress making science courses more rigorous and academic as a way to improve learning. Throughout the whole history of biology, teacher-made and standardized tests (Murnane and Raizen, 1988) have reinforced the notion that the memorization of a large technical vocabulary is equivalent to understanding biology.

There has never been a mechanism or a system developed for channeling the research on learning and cognition into the education of biology teachers, the textbooks and tests they use, and instructional procedures for making student learning more productive, in terms of knowing what it means to understand something and how to make intellectual use of it. Now that we have reached a phase in history in which there is a need for all people to be able to renew and extend their knowledge base throughout their entire life span, what is meant by knowing, understanding, and using are major components of a curriculum-implementation program.

It has been my purpose here to indicate that there is much more to a viable implementation of a reform in biology education than restructuring institutions and reformulating the curriculum, although both these endeavors are essential. As every ecologist knows, there is never an instance in which only one thing happens at a time. We would do well to think in terms of the ecology of educational reform.

REFERENCES

AAAS (American Association for the Advancement of Science). 1987. What Science is Most Worth Knowing? Washington, D.C.: AAAS.

ACS (American Chemical Society). 1988. ChemCom: Chemistry in the Community. Dubuque, Iowa: Kendall/Hunt Pub. Co.

Buccino, A., P. Evans, and G. Tressel. 1982. Science and Engineering Education: Data and Information. Washington, D.C.: National Science Foundation.

Carnegie (Carnegie Forum on Education and the Economy). 1986. A Nation Prepared: Teachers for the 21st Century. New York: Carnegie Corporation of New York.

Cole, M., and P. Griffin. 1987. Contextual Factors in Education, pp. 5-8. Madison, Wis.: Wisconsin Center for Educational Research.

CED (Committee for Economic Development). 1985. Investing in Our Children. New York: CED.

ECS (Education Commission of the States). 1983. Action for Excellence. Denver, Colo.: ECS.

Ford Foundation. 1984. City High School: A Recognition of Progress. New York: Ford Foundation.

GAO (U.S. General Accounting Office). 1984. New Directions in Federal Programs to Aid Mathematics and Science Teachers. Washington, D.C.: GAO.

Hurd, P. D. 1961. Biological Education in American Secondary Schools 1890-1960. Washington, D.C.: American Institute of Biological Sciences.

Hurd, P. D. 1969. New Directions in Teaching Secondary School Science. Chicago: Rand McNally and Co.
Hurd, P. D. 1984. Reforming Science Education: The Search for a New Vision. Washington, D.C.: Council for Basic Education.
Hurd, P. D. 1985. Science education for a new age: The reform movement. Nat. Assoc. Sec. Sch. Princ. Bull. 69:83-92.
IEA (International Association for the Evaluation of Educational Achievement). 1988. Science Achievement in Seventeen Countries: A Preliminary Report. New York: Pergamon Press.
Kirst, M. 1984. Who Controls Our Schools? Stanford, Calif.: Stanford Alumni Association.
Machiavelli, N. 1977. The Prince. J. B. Atkinson, Trans. Indianapolis, Ind.: Bobbs-Merrill Educational Publishing.
Mayer, W. V. 1986. Biology education in the United States during the twentieth century. Quart. Rev. Biol. 61:481-507.
Murnane, R. J., and S. A. Raizen, Eds. 1988. Improving Indicators of Science and Mathematics Education in Grades K-12, pp. 40-73. Washington, D.C.: National Academy Press.
NAS (National Academy of Sciences, National Academy of Engineering). 1982. Science and Mathematics in the Schools: Report of a Convocation. Washington, D.C.: National Academy Press.
NASTS (National Association for Science, Technology, Society). 1987. Bulletin of Science, Technology and Society. University Park, Pa.: STS Press.
National Commission on Excellence in Education. 1983. A Nation At Risk: The Imperative for Educational Reform. Washington, D.C.: U.S. Government Printing Office.
NEA (National Education Association). 1894. Report of the Committee of Ten, Twelve, Fifteen. New York: American Book Company.
NSB (National Science Board, Commission on Precollege Education in Mathematics, Science and Technology). 1983. Educating Americans for the 21st Century: A Report to the American People and the National Science Board. Washington, D.C.: National Science Foundation.
President's Science Advisory Committee. 1959. Education for the Age of Science. Washington, D.C.: The White House.
Raizen, S. A., and L. V. Jones, Eds. 1985. Indicators of Precollege Education in Science and Mathematics: A Preliminary Review. Washington, D.C.: National Academy Press.
Resnick, L. D. 1987. Education and Learning to Think. Washington, D.C.: National Academy Press.
Steelman, J. R. 1947. Manpower for Research. President's Scientific Research Board, Science and Public Policy. Vol. 4. Washington, D.C.: U.S. Government Printing Office.

32

Changing Practice in High Schools: A Process, Not an Event

GENE E. HALL

The 30-year period 1958-1988 has presented fantastic increases in our understanding in science. A parallel rate of increase can be documented in terms of our understanding of science education. The strategies now used to develop curriculum for the teaching of science in high schools mirror our increased sophistication in science and science education. The occurrence of a conference, such as this one, and the inclusion of topics that in many instances were unknown, or at least little understood, in 1958 are additional indicators of our learning.

In addition to a greatly increased body of knowledge, in terms of science, the importance of teacher education is now recognized. The inclusion of a panel dealing with teacher preparation and, more significantly, asking two panels to deal with institutional barriers and issues of implementation reflect major shifts in understanding, as well as significant increases in research-based knowledge. Each of these has contributed to the development of new models and strategies.

In this paper, I will describe a series of factors from studies that have documented the importance of addressing issues of implementation from the very beginning of the curriculum-development process. For example,

---

Gene E. Hall received a Ph.D. in science education in 1968 from Syracuse University. He served for 18 years at the Research and Development Center for Teacher Education at the University of Texas. He is currently dean, College of Education, University of Northern Colorado. His research emphasis is on examination of the change process from the teacher's perspective in schools and colleges.

the setting of design specifications for new curricula has direct implications for teacher training and the steps that will need to be taken to enhance the rate and ease of use of the resulting product by teachers in real classrooms. Adding to this discussion will be consideration of the unique characteristics of American high schools.

The stereotypical image of high schools is that teachers and the institution are resistant to change. In fact, Ducharme (1982) went as far as to suggest that "high schools will change when dogs learn to sing." Others suggest that high schools have not changed since the introduction of the Carnegie unit near the turn of the century. Clearly, the unique characteristics and conditions of high schools must be considered when one is thinking about strategies and ways of updating, enhancing, refining science-teaching practices.

DEVELOPMENT VS. IMPLEMENTATION

Thirty years ago, there was a singular focus on development activities when it was determined that changes in classroom practice were needed. Design teams were established that would bring together in curriculum-development projects scientists, science educators, learning theorists, and outstanding teachers. The concept of implementation was not addressed. The result was that the new curricula of the 1960s were not introduced in most of the nation's classrooms and use did not continue in most of the classrooms where they were placed. Until the count of nonadoptions soared, there seemed to be an assumption that truth (i.e., the talent in science knowledge), beauty (i.e., attractive materials), and being right (i.e., discovery approach) would automatically result in a widespread rush of regular classroom teachers to the new and dramatically different.

When the adoption rates did not increase, attempts to disseminate information about the new curricula became more systematic. At that time, the Educational Resources Information Center (ERIC) was established to handle the dissemination and adoption of new curriculum. It was not until the 1970s that there was a widespread recognition that dissemination did not necessarily lead to trial use and most certainly did not lead to continuing use of new materials. In fact, institutionalization of the many nationally developed curricula has now been well documented to be rare. One outcome of these experiences and early studies (e.g., Rogers and Shoemaker, 1971; Havelock, 1971) was the identification of a set of phases in the "knowledge utilization" process.

A major reason for the widespread nonuse of new practice was that nearly all the time, if not all the time, personnel, resources, and policy-maker attention were exhausted in the development phase. We now know that curriculum implementation requires equal time, resources, dollars, and personnel. This 1980s shift in perspective is reflected in several curriculum-development agencies, such as the Biological Sciences Curriculum Study (BSCS), and in a few school districts, such as Jefferson County, Colorado, where policies and procedures for development are paralleled by sets of policies and procedures that address implementation.

In the case of BSCS and its recent K-6 science curriculum-development project, implementation was considered a part of early curriculum development in the initial proposal, and experts in implementation were among the first personnel employed for the project. In Jefferson County schools in Colorado, a systematic set of steps have been established for the development of curriculum, including field-testing. This phase is followed by a multiyear staff-development plan to support implementation across the entire district (Melle and Pratt, 1981).

Some critics and most policy-makers refuse to consider implementation, because its cost can run high. (Others recognize that acknowledging the implementation phase would be admitting to not being able to fix things completely before the next election.) Unfortunately, we have a well-documented trail of costly curriculum-development activities that have led to no changes in classroom practices. This is especially true at the high-school level. This is not due to weak materials; rather, the failure to have widespread and sustained use of new curriculum is directly related to lack of support for teachers during the implementation phase.

We do know some things about change in high schools and effective leverage points. These factors include characteristics of the teacher, phenomena related to the innovation and its implementation, the role of the department head and the principal, and the relationship of external facilitators to implementation at the classroom level.

A basic assumption of all the research concepts that are sampled for this paper is that *change is a process, not an event.* Just as the development of a curriculum entails a process, time, a number of actors, extra resources, and iterations, so does implementation. Becoming a skilled user of a new curriculum will occur only with time. The ease of acquiring competence and confidence in use of a new way can be facilitated if various actors play their roles effectively. Unfortunately, the typical practice seems to entail many instances of activities that not only do not support, but in many cases inhibit or retard successful implementation at the classroom level.

STEREOTYPES ABOUT HIGH SCHOOLS

In one of our studies in the early 1980s (Hall et al., 1984), pairs of matched high schools were observed through the use of a "parachute drop" research technique. The schools in the sample ranged from inner-city to rural areas in the United States. The schools were paired in the sense that one school was nominated by local experts as being "more successful" with change and innovation, while the second school was considered to be less successful. The research technique entailed having two researchers on site for 2-3 days to observe school activities and interview teachers, the principal, students, secretaries, district-office personnel, and others. The goal of the study was to identify and describe factors and issues related to successful change in high schools. One immediate finding was the identification of a set of myths and stereotypes about high schools.

We identified a number of areas where the popular understanding (i.e., folklore) about high schools was in conflict with the realities that were observed by the research team. The direct implication of this study is that curriculum development needs to be based on the real world, rather than on the commonly held myths and stereotypes. The following myths and stereotypes illustrate the important point that the typical American high school is not like what many folks believe.

> If anything is certain, it is that everyone has an opinion about high schools and what should be done about them. Although each person's assumptions and assertions are somewhat unique, there are several more frequently heard stereotypes and myths that we now seriously question [Hall et al., 1984, p. 59].

Myth 1: High Schools Are Large.

The perception of high schools is one that brings with it the expectation, or the assumption, that there are many students. Typically, the image is of high schools of 2,000-3,000 and the extremes of 5,000, 6,000, or even 8,000 students. Interestingly, most high schools are relatively small. There are more high schools that have enrollments under 1,000 than over 1,000. In fact, there is a significant number of high schools (perhaps most) that have enrollments well under 500. In terms of development and use of curricula, the problem should not be considered from the point of view of high schools' being large, but rather in terms of many small high schools that are striving to deliver a comprehensive curriculum with few teachers and students and limited space and resources.

Myth 2: Principals Are Too Busy to Lead School-Improvement Projects.

Regardless of the size of the school, the principal has a very large array of responsibilities. Yet in many high schools, it has been clearly documented that the principal *can* be the instructional leader. It is probably easier and certainly safer for them to become consumed in the other tasks and responsibilities of the principalship. Yet some high-school principals do take teaching and learning seriously. They become directly involved in setting a vision and providing support for teaching. In schools that are judged as being more successful, it is clear that the principals have been key players. And these principals can be found in inner-city Philadelphia, rural Kansas, and Colorado Springs. These principals shape the context, rather than being shaped by it.

Myth 3: Cocurricular and Extracurricular Programs Are Superfluous.

It was a consistent observation and clear impression of the researchers in our studies that in schools where there was high academic success there was also extensive student involvement in a wide array of cocurricular and extracurricular programs. The two go together. Unfortunately, there appears to be no research on the role and place of cocurricular programs in the high-school setting. In the development of curricula for schools in the 1990s, it would be wise for the curriculum developers to consider the possible contributions and support that can be derived from the cocurricula.

Myth 4: You Can Find Someone Who Knows What Is Going on Inside A High School.

One of the surprises was that no one could be turned to as a source of information about all the activity that was occurring in any particular high school. This was true not only of faculty and staff, but of students as well. As we continued our field work, we kept adjusting our sampling procedures to see whether there was a role group that could give a broad-spectrum report about a school's activities. We concluded that there was no person or role group that could be relied on. In one school, a senior tight end on the football team was an amazing source of information about a wide range of activities and social traits in the school. In others, it was the secretary or a special teacher. In many, we concluded that no one, not even the principal, was aware of the total fabric. This lack of overall knowledge may be one reason for the very real isolation that most teachers, as well as students, experience in the typical high school.

Myth 5: High-School Teachers Do Not Change.

There is a widespread belief that high-school teachers are extremely resistant to change and unchanging. To our surprise, we found that most high-school teachers were interested in change; even more surprising, teachers pointed to changes that were taking place in their schools and classrooms (Rutherford and Murphy, 1985). Our conclusion was that high-school teachers are interested in improving, but the classroom, the ever-rotating 50-minute x 28-student modules, the greater school context, and external demands inhibit participation in change.

STUDIES OF CHANGE FROM THE TEACHER'S POINT OF VIEW

Thirty years ago, the talk was that curriculum development should be "teacherproof." The intent at the time was that the materials and design would be so well done that teachers would use the final product *as created*. "Teacherproof" took on a different meaning when teachers did not use these curricula. This phenomenon, probably more than any other, led to the emphasis of the 1970s on developing an understanding of implementation.

Out of the studies of implementation by my colleagues and me, two developmental dimensions were identified that describe what happens from the teachers' point of view as they become increasingly competent in using educational innovations. These two dimensions are Stages of Concern about the innovation and Levels of Use of the innovation (Hall and Hord, 1987).

Stages of Concern

The Stages-of-Concern dimension addresses the feelings and perceptions that teachers have as they are introduced to and begin to use an educational innovation. The pioneering research by Fuller (1969) on what she called the "concerns" of teachers about teaching led to a set of studies dealing with the idea that as teachers are involved in change they move through a set of stages of concern about the innovation.

The concerns theory states that at the beginning of change teachers will have more intense *self concerns* about their use of an innovation. In this stage, much thought is given to their feelings of adequacy to use the new way, their potential to be successful in using the innovation, and worries about the possible impact of the innovation on their classroom practices. They also focus on how their supervisors will judge their use of the innovation (Marsh and Jordan-Marsh, 1985).

As use of the innovation begins, *task concerns* tend to become more intense. Concerns focus on issues of time, scheduling, logistics, materials, and efficiency in using the innovation. Interestingly, this appears to be a critical time in terms of change success. If these task concerns are not addressed and resolved, the tendency of the teacher is to make changes in the innovation or to abandon use of the innovation. None of us likes to live indefinitely with intense task or self concerns, and we naturally strive to resolve these concerns, even if it means putting the innovation on the shelf.

If the task concerns are resolved and there is continued support, then it is possible that teachers will move on to having more intense *impact concerns*. At this time, teachers become concerned about how well their use of the innovation is working, in terms of effects on students, and/or they are concerned about how they can work more effectively with fellow teachers who are using the innovation.

In terms of curriculum development and implementation, the stages-of-concern dimension is one way to monitor implementation progress, as well as a framework to guide the development of dissemination strategies and implementation-support strategies. When the concerns of teachers are addressed as they arise, it is clear that implementation success is higher (Hall and Hord, 1987).

Levels of Use

A parallel phenomenon has been observed and documented relative to the *use* of an innovation. Typically, it has been thought that the teacher either uses or does not use the innovation. However, in our studies in the 1970s, it was observed that there were different *Levels of Use* (Hall et al., 1975). In terms of what teachers do with an innovation, different levels of use and nonuse could be identified. Changing practice is not an instantaneous move from doing nothing to total implementation. There is graduated iterative development of competence by each teacher individually.

At the simplest level, teachers begin at *Level O, Nonuse*. They exhibit no behaviors related to the innovation. They do not read about it or talk about it. There are no artifacts of the innovation present in their classrooms. Interestingly, in our implementation studies we have documented that surprisingly large proportions (e.g., 20%) of the teachers in "treatment groups" were not using the innovation. These individuals can have drastic effects on summative evaluation studies (Hall and Loucks, 1977).

At *Level of Use I, Orientation*, teachers are inquiring about the innovation, talking with others, reading, and studying, but have made no commitment to use the innovation. At *Level of Use II, Preparation*, there is a commitment to use the innovation, but use is not started.

Five behavioral profiles were identified relative to being a "user" of an innovation. At *Level of Use III, Mechanical Use*, teachers are working in a disjointed and relatively inefficient way with the innovation. Their planning is focused on the day-to-day activities, rather than on being able to anticipate and project what the longer-term needs and processes will require. The next level of use that was observed regularly in teachers was *IV A, Routine*. At this level, teachers have the systems worked out, they know where they are going, and they are not making adaptations in the innovation or their behaviors. They are using it in a way that has equilibrium and continuity.

For curriculum-development projects, there are several important implications of the levels-of-use concept. For example, at least 60-70% of the first-time users will be at *Level of Use III, Mechanical Use*. The longer this phase lasts, the more likely that use of the innovation will be drastically mutated or discontinued. If the curriculum materials and processes are more easily implemented or if needed staff development and coaching are provided, teachers will exhibit mechanical use behavior for a shorter period. Then they are more likely to move to *Levels of Use IV-VI*, and use of the innovation is more likely to be institutionalized.

Another implication is that summative evaluations should not be done while teachers are at the mechanical level of use. When their use is disjointed, teachers are working in ways that are less efficient than what is considered ideal. For summative evaluations, teachers should be at *Level of Use IV A, Routine*.

ADAPTATIONS IN THE INNOVATION

A major distinction should be made between characteristics inherent in the innovation (i.e., the process or product that is being implemented) and phenomena associated with the users (i.e., teachers). An important earlier study, which incidentally was done in high-school biology classrooms, was by Gallagher (1967). In this study, which I believe was the first of its kind, Gallagher documented the day-to-day practices of four teachers that were implementing the BSCS blue version. The major finding was that each of the teachers was exhibiting extremely different behaviors and practices, although all were using the *same* BSCS version! This study in and of itself should have ended the discussion of "teacherproof"; however, it seems to have taken another 10-15 years for the curriculum developers and evaluators to recognize the importance of "fidelity" as a transforming phenomenon.

In some curriculum-development projects, extreme efforts have been taken to prescribe the details of "appropriate" teacher practices and implementation requirements. In such cases as DISTAR and ECRI Reading, the developers have gone to great lengths in describing their materials and required teacher training to ensure that all teachers perform in exactly the same manner. At the other extreme, some curriculum developers, such as those of the earlier Elementary Science Study, encouraged teachers and children to "mess about"; teachers interpreted the message to mean that there was no right way. In the prior case, high fidelity is the goal that teachers can resist. In the latter case, evaluating innovation quality and effectiveness is problematic from the outset, since what "use" means is undefined.

My point here is not to advocate high or low fidelity. Rather, I am pointing out that in developing curriculum, one of the major policy decisions has to do with the degree to which fidelity of use will be expected. This is not simply a philosophical or ideological question. If the developers do not set at least minimal standards of what it means to use the innovation appropriately, it is nearly impossible to blame innovation users for failure to reach goals. If developers insist on high fidelity, implementation supporters should anticipate greater teacher resistance (i.e., higher self-concerns), because teacher autonomy is threatened.

Implementation research has made clear that addressing issues of fidelity is important at the time of development. Otherwise, evaluations become problematic and adoption of the innovation becomes increasingly less likely. In this time of accountability, it seems that more, rather than less, description of expectations about classroom practice in relation to use of innovations is needed.

THE PRINCIPAL

In earlier research, we documented a direct relationship between the intervention behaviors of the principals and teachers' success in implementation. The correlation in our principal-teacher interaction study was 0.76 (Huling et al., 1983)! Other studies in elementary schools (Venezky and Winfield, 1979; Thomas, 1978; Vandenberghe, 1988) and a major literature review (Leithwood and Montgomery, 1982) have documented the elementary-school principal's crucial role.

At the high-school level, there have been few studies of implementation. In high schools, the principal is farther removed from the classroom. There are a series of vice principals and department chairs that interface between the principal and teachers. Although the studies are fewer and less detailed, there is agreement that high-school principals are important; they should not be left out of the consideration of factors that can affect optimal success, widespread use, and quality implementation of new curricula.

THE DEPARTMENT HEAD

The department head has been a neglected factor in high schools. A surprising number of principals and superintendents are adamant that department heads should be overlooked when it comes to the change process. Others advocate that this role is key to successful implementation of new practices.

In the only major study that has been done in relation to the department head, Hord and Diaz-Ortiz (1987) note:

> It seems reasonably clear that stipends are not a critical issue for DH's. . . . What does make a difference is the allocation of time for doing the DH job.

> Having time free during the duty day to interface with teachers, to observe classroom teaching, to provide feedback to teachers, to assist them in their professional development, and to interface with administrators is an essential requirement. More active department heads do this, and they are provided the time to do it. . . . School and district administrators and policy makers will need to take these needs for time into account and supply resources accordingly, if they have interest in making the DH role viable for school improvement efforts [Hord and Diaz-Ortiz, 1987, p. 148].

It is not necessarily the contractual agreement, the official teaching load, or the type of formal training that the department head has had that makes the difference. The actual role of department heads is related to how principals treat their department heads. For example, we have observed a number of settings where department heads, by contract, were not to be involved in teacher evaluation. However, there were schools within these districts where department heads were working very effectively with an evaluation role. This was sanctioned by a principal and at least implicitly agreed to by teachers.

In terms of curriculum development, more thought should be given to ways that the department head can be used as a facilitator of implementation. The department head has a support role, even if it is only in communication and the annual allocation of budget. Where conditions permit, department heads can make a major difference in terms of the extent and quality of implementation support.

BACK TO THE HIGH SCHOOL

Attempts at schoolwide change are rare and extremely difficult to accomplish. One of the larger and more coordinated attempts to do this is that of the Coalition of Essential Schools, led by Ted Sizer. Anyone who is reading its newsletter, *Horace* (Coalition of Essential Schools, 1988), will learn firsthand of the experiences and feelings of principals, teachers, and Sizer's staff about how difficult it is to encourage and sustain change in high schools.

The difficulty, in large part, comes from within. There is a great deal of sameness about high schools across this country. When we were doing our "parachute-drop" study, we were surprised at the similarity in schools, in terms of routine, schedule, roles, organization, etc. Furthermore, it seems as though high-school programs run themselves. We suspect that you could remove one part of the school, and the rest of the school would keep running in its routine ways. In many schools, the other parts would not even notice that something was gone. Alternatively, you could insert a completely new set of players for one part of the school, and the rest of the school would continue as before.

One consequence is extreme isolation for individual teachers. It appears that many do not feel as though they are truly supported or active members of their departments (or the school). Some attempt to retain an identity through cocurricular and extracurricular activities. Others focus their psychic energies and time on other parts of their lives.

Addressing the loneliness and isolation of teachers in high schools may be a key to success in future change efforts. Absence of ways to participate in the larger community may result in forced attachment to the status quo. The often observed lack of teacher interest in students may be related to this isolation.

Perhaps some of the work by the Johnson brothers (Johnson and Johnson, 1987) and Slavin (1988) on cooperative learning groups can be applied to high-school teachers. Perhaps a curriculum should be developed that would require teachers to be supported in breaking the monastic role of self-contained classroom teacher. Perhaps the next generation of curricula should be developed not with the image of the self-contained specialists, but with an image of multidisciplinary teams of teachers working cooperatively with large groups of students who are also working cooperatively.

EXTERNAL FACILITATORS

In terms of curriculum change at the high-school level, one of the major successes of the 1960s was the program of National Science Foundation summer institutes and academic-year institutes. It is unfortunate that there has not been a systematic followup study of the many participants in these institutes. Having been an observer of many of these institutes and a trainer at several, I know firsthand of the subsequent track records and activities of many of the participants. All became more knowledgeable and skilled science teachers.

By now, many of the participants have left the classroom to become school, district, and college science educators and administrators. Those who have stayed in the classroom and the younger ones who have joined them have become increasingly out of date. This may explain why science achievement-test scores seem to be dropping. Perhaps the teachers' knowledge is not as up to date, since the resources for teacher renewal are nowhere near as extensive as they were in the 1960s.

Another element of the legacy of those institutes was the strong cadre of university science-education faculty who became active in school-level science teaching. Through their methods courses and in-service workshops, they offered teachers new ideas and advanced training. They too have had limited resources and support for renewal in the last decade.

Another external facilitator group is the science consultants in larger districts. In our district-office studies (Hall, 1987), we observed that in districts where one curriculum area, such as reading, mathematics, or science, was extraordinarily strong, there was likely to be a regionally, if not nationally, recognized curriculum coordinator.

Clinical impressions, as well as the limited amount of data that are available, suggest that these external facilitators are a key to the quality and extent of change and implementation success that are found in classrooms in high schools. Yet, today, there are limited training and support for these various external facilitators. If there is going to be widespread implementation of new programs, keys to success will be the skill and availability of external facilitators located in district offices, on university campuses, and with professional associations.

The types of training and the role expectations for external facilitators are not congruent with the way they spend their time. Van Wijlick (1987) has documented the daily intervention behaviors of external facilitators in the Netherlands. He observed that external facilitators appear to be more skilled and ready to carry out their role during the dissemination and adoption phases. During the implementation phase, external facilitators became less active, and their interventions did not shift in terms of function or target to take into account the fact that the action should now be focused on teachers. In the dissemination and adoption phases, external facilitators spent more time with the principal. However, in the implementation phase, when teachers were learning to use the innovation, external facilitators did not shift to coaching teachers; they still focused on principals! Thus, the coaching and support for implementation that teachers need (see Joyce and Showers, 1982) were not provided.

In the next round of curriculum development, thought needs to be given not only to who the external facilitators will be, but to what their roles should be during different phases of the change process. Furthermore, the Van Wijlick study (1987) suggests that training for the coaching role that is so critical in the implementation phase may need to be provided for external facilitators. They do not appear to be skilled in what their role needs to be during the implementation phase, or for some reason they are not predisposed to perform this role.

SUMMARY AND RECOMMENDATIONS

There appears to be a natural evolution to the focus that policy-makers give to strategies and approaches to change for curriculum in the United States. In the 1960s, the emphasis was on the development of "teacher-proof" curriculum. In the 1970s, the vogue was "mutual adaptation." In the early 1980s, the focus was on the "cottage industry," in which teachers in each school would develop their own curricula.

We are headed toward a time of joint or collaborative ventures. Cooperative, multidisciplinary, international, and integrated activities are all keys to the development of curriculum for the 1990s. This approach is reflected in the BSCS K-6 elementary-school science program. In this effort, science educators, teachers, publishers, experts in technology, experts in health education, *and* experts in implementation have joined together to develop curriculum and to plan from the very beginning for dissemination and implementation support.

A major key for achieving implementation success across this country is to include, from the very beginning of discussions of curriculum development, change-process experts. Someone who knows the research, theory, and practical aspects of dissemination, implementation, and institutionalization should be involved from the outset. These individuals can be of great help in designing evaluation procedures. Their expertise will be helpful in anticipating problems that may retard success. Also, by being knowledgeable from the very beginning in the philosophy and intent of the curriculum developers, dissemination and implementation experts can better plan for the marketing activities and strategies to promote wide-scale use of the resulting product.

The *fidelity* issue needs to be addressed at the beginning also. To what extent will teachers be encouraged to be creative and adapting? Are there certain elements or components of a program that *must* be adhered to? Which outcomes need to be emphasized most? A useful distinction is the one between implementation requirements and descriptions of ideal use of an innovation. Specifying that there should be 90 minutes for a laboratory activity is an implementation requirement. Describing what teachers and students should be *doing* during that 90 minutes addresses the idea of "key elements" and configurations of use (Hall and Loucks, 1981). Also, it is important that the decisions about fidelity be addressed and arrived at collaboratively. Curriculum developers used to try decrees; in the 1970s, state policy-makers decreed. As we head toward the 1990s, it will be important to involve all role groups and to be able to explain to adopting teachers the reasoning for certain curriculum directions.

External facilitators are essential to any attempt to have widespread use of new curriculum materials in high schools. The long-term dominance of the textbook market by *Modern Biology* is a clue that bringing about widespread change in practice will not be easy. Change will require taking into account the contexts of high schools, the role of external facilitators, the characteristics of the innovation, and the needs of teachers and students; the department head may be the key.

Change in high schools is a process, not an event. Just as the development of the curriculum takes time, energy, and resources, so too will development of the specialized skills and resources that are needed to succeed with implementation. A carefully developed strategic plan (including a marketing plan) will be needed to support and facilitate teachers in a multiyear effort to master new ways.

It can be done; however, to do it well will require an effort that takes into account implementation and institutionalization, as well as the many requirements of curriculum development. Since change is a process, there needs to be time to grow with learning to competence and sophistication. Developing competence and confidence in doing new things requires continuity, stability, personalized coaching, and time.

REFERENCES

Coalition of Essential Schools. 1988. Horace 4(3).

Ducharme, E. R. 1982. When dogs sing. The Prospect for change in American high schools. J. Teach. Educ. 33:25-29.

Fuller, F. F. 1969. Concerns of teachers: A developmental conceptualization. Amer. Educ. Res. J. 6:207-226.

Gallagher, J. J. 1967. Teacher variation in concept presentation in BSCS curriculum programs. *BSCS Newsl.* 30:8-19.

Hall, G. E. 1987. The role of district office personnel in facilitating school-based change: Hypotheses and research dilemmas, pp. 163-184. In Research on Internal Change Facilitation in Schools. Leuven (Belgium): ACCO (Academic Publishing Company).

Hall, G. E., and S. M. Hord. 1987. Change in Schools: Facilitating the Process. Albany, N.Y.: State University of New York Press.

Hall, G. E., and S. F. Loucks. 1977. A developmental model for determining whether the treatment is actually implemented. Amer. Educ. Res. J. 14:263-276.

Hall, G. E., and S. F. Loucks. 1981. Program definition and adaptation: Implications for inservice. J. Res. Dev. Educ. 14(2):46-58.

Hall, G. E., S. F. Loucks, W. L. Rutherford, and B. W. Newlove. 1975. Levels of use of the innovation: A framework for analyzing innovation adoption. J. Teach. Educ. 26(1):52-56.

Hall, G. E., S. M. Hord, W. L. Rutherford, and L. L. Huling. 1984. Change in high schools: Rolling stones or asleep at the wheel? Educ. Leader. 45:58-62.

Havelock, R. G. 1971. Planning for Innovation Through Dissemination and Utilization of Knowledge. Ann Arbor, Mich.: University of Michigan, Institute for Social Research.

Hord, S. H., and E. M. Diaz-Ortiz. 1987. Beyond the principal: Can the department head supply leadership for change in high schools, pp. 117-154. In Research on Internal Change Facilitation in Schools. Leuven (Belgium): ACCO (Academic Publishing Company).

Huling, L. L., G. E. Hall, S. M. Hord, and W. L. Rutherford. 1983. A Multidimensional Approach for Assessing Implementation Success. Report No. 3157. Austin, Texas: University of Texas at Austin, Research and Development Center for Teacher Education. (ERIC Document Reproduction Service No. ED 250-328)

Johnson, D. W., and R. T. Johnson. 1987. Learning Together Alone. 2nd ed. Englewood Cliffs, N.J.: Prentice Hall.

Joyce, B. R., and B. Showers. 1982. The coaching of teaching. Educ. Leader. 40:4-10.

Leithwood, K. A., and D. Montgomery. 1982. The role of the elementary school principal in program improvement. Rev. Educ. Res. 52:309-339.

Marsh, D. D., and M. Jordan-Marsh. 1985. Addressing Teachers' Personal Concerns in Staff Development Efforts. Paper presented at the annual meeting of the American Education Research Association, Chicago.

Melle, M., and H. Pratt. 1981. Documenting Program Adaptation in a District-wide Implementation Effort: The Three-Year Evolution from Evaluation to an Instructional Improvement Plan. Paper presented at the annual meeting of the American Educational Research Association, Los Angeles.

Rogers, E. M., and F. F. Shoemaker. 1971. Communication of Innovations: A Cross Cultural Approach. 2nd ed. New York: Free Press.

Rutherford, W. L., and S. C. Murphy. 1985. Change in High Schools: Roles and Reactions of Teachers. Paper presented at the annual meeting of the American Educational Research Association, Chicago.

Slavin, R. E. 1988. Cooperative learning and student achievement. Educ. Leader. 46(2):31-33.

Thomas, M. A. 1978. A Study of Alternatives in American Education. Vol. II: The Role of the Principal. Santa Monica, Calif.: Rand Corp.

Vandenberghe, R. 1988. The principal as maker of a local innovation policy: Linking research to practice. J. Res. Devel. Educ. 22(1):69-79.

Van Wijlick, W. 1987. The activities of internal degree change facilitators: An analysis of interventions, pp. 99-116. Research on Internal Change Facilitation in Schools. Leuven (Belgium): ACCO (Academic Publishing Company).

Venezky, R., and L. Winfield. 1979. Schools That Succeed Beyond Expectations in Teaching Reading. Newark, Del.: University of Delaware.

33

Change in Schools: A Context for Action

DEBORAH MUSCELLA

A SYSTEM PERSPECTIVE OF SCHOOLS

What propositions about change in schools should we test as we introduce science curricula into high schools? There is considerable research regarding change in schools. Researchers have considered the school as a culture, a social ecology, and a system of people who act both individually and collectively (Berman and McLaughlin, 1974; Erickson, 1986; Fullan, 1982; Hall and Hord, 1986; Lortie, 1975; Moos, 1985; Saranson, 1971). One theme central to this research is that change is both psychological and sociological. Put more simply, for change to occur in schools, someone has to change. In their efforts to change, however, people are influenced by the contexts in which they live and work.

Schools are part of a social matrix; the many facets of this matrix influence one another. The community and the school district affect what happens in the school and the classroom, and these environmental influences affect what teachers, principals, and students do. When a new curriculum is introduced into a school, the components of the system will

Deborah Muscella is the implementation specialist for the elementary-school science curriculum that the Biological Sciences Curriculum Study is developing. She received her Ph.D. from The University of Texas, Austin, where she worked as a research fellow at the Research and Development Center for Teacher Education. She has conducted research to examine school change and has consulted with teachers and administrators to assist them in their school-improvement efforts.

affect its implementation. A systems perspective of the school is a necessary framework for understanding what is required for curriculum renewal in high-school biology.

THE ECOLOGY OF A LEARNING ENVIRONMENT

Learning environments are multidimensional. Thus, there is not a simple causal connection between learning and teaching. Rather, a variety of factors promote or inhibit what students learn. There are the curricular materials that come to life through the particular pedagogy that the teacher employs. Knowledge, previous learning experiences, and cultural information influence the way in which students participate in pedagogy (Erickson, 1986; Ittelson et al., 1976; Lortie, 1975; Muscella, 1987; Novak and Gowin, 1984). The context of the classroom, which has established rules and norms for the interaction of instruction, affects how teachers and students immerse themselves in a new curriculum (Erickson, 1986; Saranson, 1972).

Schwab (1973) deems these the four commonplaces of the classroom: teachers, students, the curriculum, and the learning environment. They are all parts of an interactive system. When other influences are added to a learning system, the complexity increases. Beyond the classroom is the school organization that has prescribed the ways and means of instruction. Not only is there a rubric for instruction, but in the school organization there are cultural nuances that influence how teachers and students interact with one another and how the staff members work together. These interactions and the framework for the instructional program of the school determine how a school adapts to change.

In a school with detailed rules for instruction and behavior, substantial changes in curriculum may be difficult to achieve. We know that, because of their beliefs, some communities do not support the teaching of evolution in high-school biology. What happens in schools is directly related to societal influences.

A social-ecological model is one way to explain the complexity of learning environments. This model provides a framework for examining how people shape and are shaped by their environment; a systems model also represents the interactions between persons and their situations (Barker, 1963; Bronfenbrenner, 1979; J. Kelly, 1969; Moos, 1979, 1985; Muscella, 1987; Nystedt, 1983). Bronfenbrenner (1979) considers the social system as a hierarchy of subsystems that impinge on the life of the individual. He proposes that to study what individuals do and yet ignore the impact of the home, community, and school gives a constricted view of the individual. Moos (1979) isolated four factors—the people, the physical environment, the organization, and the social climate—that are the critical elements to consider in environments. Nystedt (1983) uses "the person-situation interaction" to describe how beliefs influence the actions that individuals take: both actions and beliefs are constrained or expanded by the situations in which people live and work. Thus, in any study of a social environment, we must consider a range of activities and needs that meet diverse functions in an organization (J. Kelly, 1969).

Considering the learning environment as an ecology is vital if we are to capture a holistic perspective of the events that constitute learning. The ecology of the school has several components—students, teachers, curriculum, administrators, environment, and higher education—and these interact with one another in a variety of ways. Because these components interact, intervening in one part of the system has consequences for other parts of the system. An intervention directed to one component may require interventions in other parts of the system.

Consider the path that a particular intervention in the renewal of the biology curriculum in the high school requires. Once the academic community has established guidelines for curricular design and developed prototypes and programs, other interventions in the system are mandatory for implementation. Before teachers can begin to teach the program, they need training in its content and pedagogy. Equally important, the principal, science supervisor, and department head must provide leadership, if the curriculum is to become integral to the school's science program. The leadership team also gives attention to adapting and modifying structures in the environment that are necessary to support the curriculum. It is the interaction among the four components of the system—administrators, teachers, the environment, and higher education—that is the critical infrastructure for curriculum renewal in biology education.

CHANGE IS SYSTEMIC

For any change in the curriculum to be meaningful and lasting, systemic changes are required; that is, changes in which all the parts of the system work in concert (Fullan, 1982; Hall and Hord, 1986; Joyce and Showers, 1988). Let us consider an interdisciplinary science curriculum in biology. What prescriptions are mandatory to ensure that the curriculum is implemented effectively?

- Key leaders work together to plan, implement, and monitor the curriculum.
- Attention to the structure of the environment is central to any implementation strategy.
- Teachers alter their beliefs about teaching and learning as a result of their participation in the pedagogy of the curriculum.

When the school is considered in a systems perspective, needed interventions can be identified. Change requires a *change agent*, a *change facilitator*, or a *leader for change* who is pivotal in establishing change as necessary and vital for the system. Because change is a process, the leaders of change give time and attention to the various parts of the system and assist and support teachers in their efforts to change. Innovations in any curriculum require changes in the beliefs that teachers hold about the teaching and learning process. For teachers to alter their role perspective, they must understand and internalize the knowledge and values espoused by a particular curriculum. For lasting change to occur, individuals in the various components of a system must lead in the effort. We will consider the enactment of change in the context of Southside High School—the idealized ecology in which curriculum renewal comes to life.

Change agents adopt a game plan and then proceed to develop strategies so that people adopt the innovation (Rogers and Shoemaker, 1971). Change agents are rarely members of the community for which a particular innovation is targeted. Therefore, one of the first things that a change agent does is to convince people of the need for change. Convincing others to change requires that the change agent convince the leaders in the community that a particular innovation will benefit the members of that community. The mark of a change agent's success is when the clients have adopted and are using the innovation and no longer require support or guidance.

A facilitator of change, as a member of the community for which the innovation is targeted, takes many actions to ensure that people adopt and use a particular innovation (Hall and Hord, 1986). In a school setting, the facilitators of change—most often the principal and the department head—enlist the support of a small cadre of teachers, and together they act as a team to facilitate change. The role that each person assumes is critical for any change to occur: the principal, as the leader for change, sets the tempo, engenders a vision for full implementation, and gives consistent time and attention to the innovation; the department chair gives the day-to-day attention that any innovation requires; teachers provide technical assistance and moral support to their colleagues. Together the team members take many actions to support their colleagues as they begin to use the innovation.

At Southside High School, who are the change agents, and how does the leadership team facilitate change? Five years ago, the biology department at the local university developed a training program for preparing teachers and administrators to use an interdisciplinary science curriculum. The state science supervisor and a school-district science supervisor attended the seminar, and their conviction that the program was vital led them to propose to the district superintendent that the school district adopt it. With the district superintendent, they developed a plan to implement the program over a 5-year period. They formed a troika with the university: together they acted as the change agents. They enlisted key informants from two high schools; these principals and science-department chairs were well regarded by their contemporaries and thus would become a necessary link in establishing support for the curriculum. To broaden the support base, a biology teacher and chemistry teacher from each school joined the team.

Notice that the change agents have given careful attention to three components of the system—higher education, administrators, and teachers. Not only did they acknowledge these components; they understood the interactions among them and involved leaders from each subsystem from the inception of curricular implementation.

Each school team then attended the training institute; and when they returned to their school district, the district and state supervisors consulted with each team and assisted them in developing a plan to implement the program in their school. The supervisors convinced the school teams that staff development would be the cornerstone of their plan for change and that such development was essential during the 3-year implementation phase. These supervisors met the school teams several times during the first year of implementation; by the third year, they had relinquished their role as change agents to the school teams.

The change agents in this story had a systems perspective of schools. In their efforts to effect change, they considered several levels of the system. Recall that they first received the approval of the district superintendent. They enlisted the support of leaders—the principal, science-department chair, and teachers—people at different levels of the school who became critical in persuading other teachers to adopt the innovation. Finally, to ensure that other levels of the school system—the district and the state—influenced the adoption and implementation of the interdisciplinary curriculum, they maintained an active role in implementation.

Change Is a Process

"Change is a process, not an event," is the signature of research on change (Hall and Hord, 1986). Researchers who have studied how change occurs in schools and in other environments document that change requires time, because there are many facets to the process (Fullan, 1982; Moos, 1985). Change is at once a psychological, sociological, and cultural process (Lortie, 1975; Moos, 1985; Saranson, 1971). For any change to take place in a school, people must change. When confronted with the task of implementing a new curriculum, educators must determine how they will use the new program effectively. Their adjustment to adoption and implementation is hierarchical, with attention first to concerns about how a new program will affect them personally until eventually teachers consider how the new program influences the learning process (Hall et al., 1984).

Although change is psychological, it takes place in a social context. Teachers extend or retract their efforts to change as they read information from this social context (Lortie, 1975; Moos, 1985; Saranson, 1971). The local norms of a setting—the tacit rules that guide collegial relationships and interactions between students and teachers—affect the commitment to implementation.

The success of the new science curriculum at Southside High School was orchestrated by the leaders for change. They targeted their interventions at critical points of the environment. Many actions were required on the part of each member of the leadership team. At a faculty meeting, the biology-department head set up an experiment that measured pulse rate with a microcomputer-based laboratory. The teachers then tested their own pulse rates. During the next few months, computers would appear in the workroom, department office, and teachers lounge. The teachers tried out simulations, problem-solving, and graphics programs. Next, the school leaders asked four teachers to volunteer to attend four seminars in which they would learn about technology in the science classroom. It was at this point that the principal arranged to have all the equipment installed in the science laboratory. When members of the leadership team had completed the seminar, they enlisted teachers to design the laboratory that would house the computer stations. In addition, four teachers agreed to consult with a team of three teachers; the department head met with these teachers each month.

Figure 1 shows how the various parts of the system are interrelated. Because these elements are nested, what happens in the classroom is affected by the community. And the education that students receive in classrooms eventually influences the larger community. The principal's attention to the systemic aspects of change prompted her to focus on processes inherent in effecting change. She enlisted the support of another change facilitator—the department head—to generate an awareness of the technology. She then probed the classroom and school-district levels by enlisting the participation of teachers and using training offered by the district. She expanded the cadre of supporters for the innovation by encouraging teachers to work in teams and establishing a monitoring system through these teams and the department head. There was evidence that the innovation was supported at the district, school, and department levels.

BELIEFS ABOUT TEACHING AND LEARNING

Innovative curricula require a change in the perspective that teachers, principals, and department heads have about their roles with their students

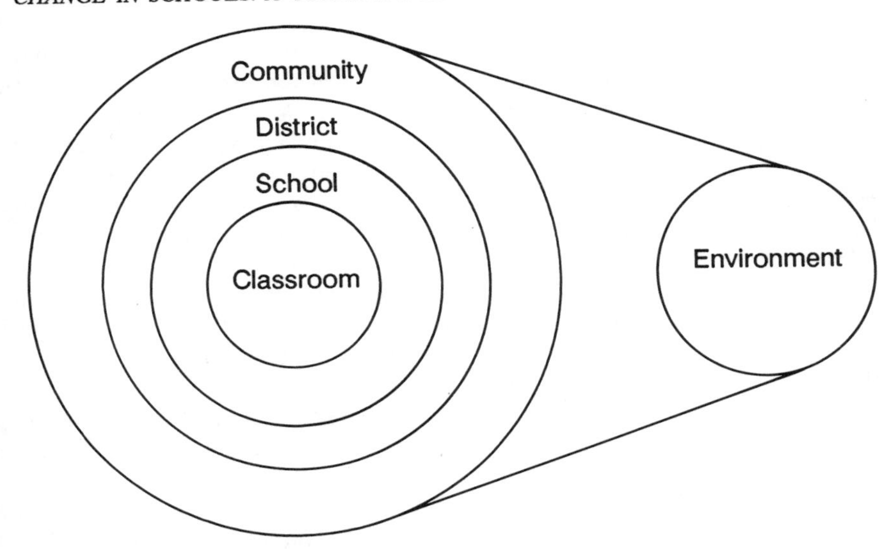

FIGURE 1 The components of the learning environment.

and with each other. Consider the models of teaching introduced into schools in the past, such as an inquiry approach to learning (Joyce and Weil, 1986; Joyce and Showers, 1988; Schwab, 1973). In this approach, the skillful teacher asks questions and directs students to new lines of inquiry. For teachers to change their perspectives on their roles, they must internalize the knowledge and values espoused in the curriculum. As Fullan and Pomfret (1977) state, teachers must have a "deep understanding" of the curriculum. It may well be that an inquiry approach to teaching biology is seldom used in high-school biology classes today because teachers had little opportunity to reconsider their roles as pedagogues.

Teachers' changing of their views about what constitutes teaching and learning is not a solitary enterprise. Indeed, for any teacher to change the fundamental way that he or she interacts with students, changes in the organization are required (Lortie, 1975; Saranson, 1971). Unless the principal and the department head sanction the changes in pedagogy required of the new curricula, it is unlikely that teachers will change. The sanction required is more than verbal approval. Rather, the principal and the department head must take specific actions to support teachers in their change efforts (Hall and Hord, 1986). These actions include training in the new pedagogy, enabling teachers to work together, and continuing consultation that is directed at the innovation (Joyce and Showers, 1988). Equally important, teachers must have time to experiment with new methods of teaching.

Several layers of the environmental system ultimately influence what teachers believe about learning environments. When teachers begin to use a new curriculum, they are mechanical, because they are learning how to maneuver among the materials, content, and pedagogy. As they become facile with these new skills, their knowledge about the program increases. As their knowledge grows and the teachers become more reflective about how the pedagogy influences what students learn, teachers begin to reconsider their roles. Often, if there is support for change in the social matrix of the school, teachers will change their role perspective. As teachers begin to move from a pedagogy that is primarily lecture and discussion into one in which they are facilitators and inquirers, they may change the beliefs that they hold about teaching and learning. It is unlikely that teachers will change their beliefs about learning unless there are structural changes in the school, district, and community environment that will support alterations in the perspectives that teachers have about their roles. The leadership team at Southside High School recognized during their own training in the curriculum that their role as teachers would change substantially. They understood that teachers rarely had training in facilitating groups; instead, their training in pedagogy focused on a lecture-discussion model. They anticipated many problems with students and parents and their other colleagues, because teachers would not have the necessary skills. The principal addressed their concerns by taking the following actions. She recommended that the leadership team spend the first year in planning and training. She also arranged for the leadership team to visit the school in a nearby community that recently had implemented the curriculum. The principal also arranged training for the teachers in group-process skills through the local university.

When teachers began to implement the curriculum, each teaching team met once a week with a member of the leadership team to discuss problems. The department head and the principal arranged staff development as the particular needs of teachers dictated. Even with these strategies to support teachers as they altered their role perspective, several teachers remained uncomfortable with the new pedagogy. It was only through the continued persistence of the department chair and other members of the leadership team that these teachers continued to use the curriculum. The leadership team was encouraged with the implementation of the program, because most teachers had changed their perspectives about their roles as teachers. The principal and the department chair believed that teachers valued the new curriculum because student achievement had improved.

Teachers may find it most difficult to change their beliefs because it is not easy to change fundamental perspectives about learning (G. Kelly, 1955). Argyris and his colleagues (1985) contend that, for people to change the way that they think about a social setting or their role in a work environment, they must learn how to consider alternatives. Most individuals have one system of constructs that they use to understand and take actions in situations—a closed system (Argyris et al., 1985; G. Kelly, 1955; Schon, 1982). People must adopt an open system, if they are to change the beliefs that prompt actions (Argyris et al., 1985). When individuals use an open system, they consider information that is discrepant from their beliefs. They accept that cognitive dissonance is essential for change to occur (Cofer and Appley, 1966).

Because beliefs and the social context are interwoven, constraints in situations often prevent people from changing their minds. A corollary is that, if there is planned change that will require individuals to change their beliefs, the social context must provide support. There is other evidence that suggests that teachers do not change their beliefs or attitudes until they have facility with new curricula (Fullan and Pomfret, 1977; Guskey, 1985). Thus, the agents, leaders, and facilitators of change must provide many and varied supports in the system, so that teachers experience success.

Many teachers at Southside High School altered their perspective about teaching, because the new curriculum was effective with students. The leadership team took many actions to make change happen. Three aspects are central to their interventions: staff development was a continuing part of the change process; teachers worked in teams to plan and learn about aspects of the new curriculum; the district, school, and community were involved in the process of change. Change was considered important, because the school leaders gave much attention to its implementation and invervened in the four major parts of the system. It is unlikely that the science curricula would have become integral to the school if the leaders had not provided a social context that supported change.

THE HIGH SCHOOL AS A CONTEXT FOR ACTION

Eisner (1988) likens a day in the life of a high-school student to a day on Broadway, in which students visit one theatrical production after another, each with a different cast and director. A student might begin with a discussion of Shakespeare during English class, solve quadratic equations in algebra class, then discuss the War of 1812, then translate a literature passage, and end the day learning about vectors in a physics class. When the day ends, students will have made few, if any, connections among the conceptual themes they encountered in their classes. In fact, because teachers do not typically plan or consult with their colleagues, the classes rarely relate to one another. This is not to say that there are not certain teachers who inspire students, convey important concepts, or evoke new ways of thinking. Rather, the 50-minute cycle that is typical of the instructional program in high schools sets a tempo in the high school that prevents connections between the concepts taught in the disciplines, isolates students, and insulates teachers. In this context, it is doubtful that students will see the connections between what they learn in school and the world they call real life, and it is into this context that we attempt to infuse new science curricula. If these curricula are to represent the reforms advocated in the many national reports, then the very context of the high school will be changed (Boyer, 1983; Sizer, 1984).

Consider an interdisciplinary science curriculum that integrates the major ideas of biology, chemistry, earth science, and physics. Consider also that the schedule of the day becomes modular, so that a science class has a 2-hour block of time. Technology will be integral to the science classes, with students using microcomputer-based laboratories, interactive video disks, and computer databases and simulations. The teachers will present new concepts and skills through laboratory experiments, lectures, discussion, and consultations.

Implicit in an interdisciplinary science curriculum is the responsibility of leadership teams to monitor and support implementation strategies, of teachers to develop new pedagogical skills and to work closely with their colleagues, and of students to assume an active role in their own learning. For curriculum renewal in biology to become a reality, leaders from all components of the system must work together throughout all phases of development and implementation. It is critical that the academic community, business, and industry work with school administrators and teachers in developing the blueprint for curriculum renewal in education. It is equally important that key leaders from the school, district, community, and higher-education arenas remain involved throughout curriculum development and implementation. Curriculum renewal will require many interventions in the various components of the system:

- Leaders from state education offices, school districts, and schools must plan, design, and implement the curriculum together.
- Leaders in the change process must modify and adapt the environmental structures in the school to support the curriculum.
- Teachers must assume an active role in implementation by working in teams with their colleagues and members of the school leadership team.

If these conditions for change are met, then lasting change in biology curricula is possible.

REFERENCES

Argyris, C., R. Putman, and D. McLain-Smith. 1985. Action Science: Concepts, Methods, and Skills for Research and Intervention. San Francisco: Jossey-Bass.

Barker, R. G. 1963. The Stream of Behavior. New York: Appleton-Century Crofts.

Berman, P., and M. W. McLaughlin. 1974. A Model for Educational Change. Rand Change Agent Study—Federal Programs Supporting Educational Change. Vol. I. R158911-HEW. Washington, D.C.: Department of Health, Education, and Welfare.
Boyer, E. L. 1983. High School. A Report on Secondary Education in America. New York: Harper & Row Publishers.
Bronfenbrenner, U. 1979. The Ecology of Human Development: Experiments by Nature and Design. Cambridge, Mass.: Harvard University Press.
Cofer, C. N., and M. H. Appley. 1966. Motivation: Theory and Research. New York: John Wiley & Sons.
Eisner, E. 1988. The ecology of school improvement. Educ. Leader. 45(5):24-29.
Erickson, F. 1986. Tasks in times: Objects of study in a natural history of teaching, pp. 131-148. In K. Zumwalt, Ed. Improving Teaching. Alexandria, Va.: Association for Supervision and Curriculum Development.
Fullan, M. 1982. The Meaning of Educational Change. New York: Teachers College Press, Columbia University.
Fullan, M., and A. Pomfret. 1977. Researchers curriculum and instruction implementation. Rev. Educ. Res. 4:335-393.
Guskey, T. R. 1985. Staff development and teacher change. Educ. Leader. 42(7):57-60.
Hall, G. E., and S. M. Hord. 1986. Change in Schools: Facilitating the Process. Albany, N.Y.: State University of New York Press.
Hall, G. E., S. M. Hord, W. L. Rutherford, and L. L. Huling. 1984. Change in high schools: Rolling stones or asleep at the wheel? Educ. Leader. 41(5):22-29.
Ittelson, W. H., L. G. Rivlin, and H. M. Proshansky. 1976. The use of behavioral maps in environmental psychology, pp. 340-351. In H. M. Proshansky, W. H. Ittelson, and L. G. Rivlin, Eds. Environmental Psychology: People and Their Physical Settings. 2nd ed. New York: Holt, Rinehart & Winston.
Joyce, B., and B. Showers. 1988. Student Achievement Through Staff Development. New York: Longman.
Joyce, B., and M. Weil. 1986. Models of Teaching. Englewood Ciffs, N.J.: Prentice-Hall, Inc.
Kelly, G. A. 1955. The Psychology of Personal Constructs. Vols. 1 and 2. New York: Norton.
Kelly, J. G. 1969. Naturalistic observations in contrasting social environments, pp. 183-199. In E. P. Willems and H. L. Raush, Eds. Naturalistic Viewpoints in Psychological Research. New York: Holt, Rinehard & Winston.
Lortie, D. C. 1975. School Teacher: A Sociological Study. Chicago: University of Chicago Press.
Moos, R. H. 1979. Evaluating Educational Environments. San Francisco: Jossey-Bass.
Moos, R. H. 1985. Learning environments in context: Links between school, work, and family settings, pp. 47-59. In B. Fraser, Ed. The Study of Learning Environments. Salem, Ore.: Assessment Research.
Muscella, D. 1987. Uncovering Beliefs about Learning: Multi-method, Multi-trait Research. Paper presented at the annual meeting of the American Educational Research Association, Washington, D.C.
Novak, J. D., and D. B. Gowin. 1984. Learning How to Learn. Cambridge, England: Press Syndicate of the University of Cambridge.
Nystedt, L. 1983. The situation: A constructivist approach, pp. 93-114. In J. Adams-Webber and J. C. Mancusco, Eds. Applications of Personal Construct Theory. Toronto: Academic Press.
Rogers, E. M., and F. F. Shoemaker. 1971. Communication of Innovations: A Cross-Cultural Approach. New York: Free Press.
Saranson, S. 1971. The Culture of the School and the Problem of Change. Boston: Allyn & Bacon.
Schon, D. A. 1982. The Reflective Practitioner. New York: Basic Books, Inc.
Schwab, J. J. 1973. The practical 3: Translation in curriculum. School Rev. 81:501-522.
Sizer, T. 1984. Horace's Compromise. New York: Houghton-Mifflin.

34
Creating and Nurturing Curriculum Changes: Some Models that Speak to the Future

FRANCIS M. POTTENGER III

As we think about the biology curriculum of the schools as it may exist in the last decade of this century and the first of the next, we have to face up to the fact that its quality will depend more on the structure of programs of dissemination and implementation than on the structure and content of materials. American science education is already at a point where, overall, high-school science teaching staffs must be rated as marginal, and there is little prospect of our turning this reality around in the near future. It follows that the real classroom delivery of curriculum to students over the next 20 years will be determined far more by the role that the curriculum disseminator takes than by the creativity of the curriculum designer. Whether the marginal teacher will be able to deliver the program that our students need will ride on whether the disseminator acts only as a persuasive broker of ideas, goes beyond and trains the teacher in the use of the materials, or goes farther still and takes on the role of the mentor-coach, helping the teacher to master both content and pedagogy of the program.

On the premise that the fate of America's biology education is tied to

Francis M. Pottenger III is professor of education in the College of Education and director of science projects at the Curriculum Research and Development Group of the University of Hawaii. He is director of the Developmental Approaches in Science and Health (DASH) project for grades K-6 and has directed development of Foundational Approaches in Science Teaching (FAST), the Hawaii Marine Science Studies (HMSS) project, and numerous other science, health, and social-studies curriculum projects.

science education generally, the next few pages describe some of the problems that we face in the curricular change process in American education and in science in particular. Then will come a discussion of the experience of the Curriculum Research and Development Group of the University of Hawaii as it has tried to come to grips with problems of long-term curricular planning, development, dissemination, and institutionalization.

CURRICULAR CHANGE

To this point in history, possibly the greatest strengths in American precollege education are its diversity and diffused control. These have given us the flexibility we have needed to provide the laborers, farmers, merchants, engineers, inventors, and statesmen for a vast and growing nation. It is this diversity that has provided rich educational experimentation; a shield against total embrace of single philosophical, social, or pedagogic panaceas; and examples of stellar educational success and dismal failure.

Change As an End

Out of our diversity and faith in the capacity of education to provide social flexibility and growth, we have made the school curriculum carry an ever-increasing load of social responsibility as expressed in mandated courses and requirements. Whenever there is a new social problem or an emerging area of technical advance, it is soon reflected in changes in the curriculum.

Over the last 30 years, we have turned curricular change into an artform. Today, every special-interest group, government agency, publisher, school board member, district specialist, principal, or teacher with a cause, a penchant, or something new to sell has a large list of rationales for undertaking curricular change and a toolkit full of ploys to bring it about. Change in curriculum has taken on the mantle of newness. So intoxicating is newness that change has become a community value, a raison d'être, an end in itself.

The School's Mechanisms for Change

In the face of a host of curricular requirements and options, a school can do one of two things. First, it can have teachers create a curriculum. Without time for design and development, this will normally take the shape of off-the-shelf materials and become a collection of units, fed by an encyclopedic text and an array of supplementary, unconnected modules used to reflect mandated requirements. In the process, mandates are honored, special interests are served, flexibility is maintained, and, most important, the claim can be made that curricular change has been effected.

Second, the school may turn to a prefabricated curriculum. This is often a necessity, if a new approach to classroom delivery is required, such as the inclusion of laboratory activities. In adopting a new curriculum, special requirements may be satisfied, but newness does not guarantee instant achievement of intended goals or instant facility in new teaching techniques. New content and pedagogic approaches require time to master and personalize, and the slowness and expense of the process tend to create frustration in both teachers and administrators. Inexorably, the prefabricated curriculum gives way to the lure of "newerness," and the cycle of change is set in motion again.

Stability in Curriculum

In many places, educators have become so enamored of change that they overlook the value of curricular stability. A curriculum that can be maintained for a period of years is usually a curriculum that is delivered with increasing skill, competence, and satisfaction. Quality curriculum delivery is produced out of long refinement.

It is much easier and less expensive to build a school program around a stable curriculum than around a curriculum in a state of flux. For the teacher dealing with 150-180 students a day, under the pressures of correcting endless stacks of homework and weekly tests, as well as an extracurricular assignment, curriculum stability makes the job possible. For the administrator, stability in curriculum is the oil that makes the school run smoothly.

However, stabilized curricular offerings have problems. Predictably, many teachers become bored with the same routine, and interaction with their classes becomes mechanical and uninspired. Students are quick to note this and respond in kind. In addition to inducing boredom, curriculum that has been in place without modification for a period of years has a high probability of becoming out of date in terms of both content and pedagogy.

Crisis, Stability, and Change

Forces for stability and change pull the curriculum in opposite directions. Both must operate, if we are to achieve and maintain educational excellence. Commentators tell us that both stabilizing and change factors should be continually weighed and programs should be monitored internally and externally to determine whether modification is warranted. At the school level, mandates from legislative bodies should be carefully analyzed for their connection with the existing curriculum, and, where possible, modification of existing programs should be nondisruptive. The emphasis of curriculum delivery should be on development of ever-increasing quality. Major change in curriculum should be undertaken with equal deliberation. Training should be given to all parties during the time of implementation of new curricula, and the goal of implementation should be to reach a state of dynamic stability as soon as possible.

Politicians choose to move education by declaration of crisis. Education and its problems seem to gain prominence in the minds of our national political leadership once every 15-20 years, at which time a national crisis is pronounced with great ceremony and the machinery of federal legislation is put into gear to remedy the newly discovered deficiencies. Once this is done, the educational establishment is expected to work hand in hand with new assistance programs generated during this time of focus and, as a good physician would do, go and heal itself.

State legislatures operate in like manner. Moved by federal concern, they fall in line, picking up and brokering new federally funded programs and adding their vision of the solution to the crisis through new curriculum guides, graduation tests, and mandated program.

Crisis once declared is infectious. Overnight, publishers, entrepreneurs, universities, and professional and special-interest groups gather round with their packaged versions of solutions.

The Current Crisis

Let us single out science for a closer study of projected response to our latest crisis, which in sum calls for increased quantity and quality of laboratory science and technology education for all students. The change levers of the federal and state governments, special-interest groups, and entrepreneurs are now going into place. But what can we expect of the efforts? What are the forces resisting change?

Teachers and Curricular Change

Although the actions of government, boards of education, and administrators are essential in the process of curricular change, it is what happens in the classroom that determines the success of curricular change. The real determinant of success is how well the teacher's needs and problems in delivering the curriculum are understood and accommodated. For the teacher facing a new curriculum with a laboratory component and a new pedagogy of cooperative learning or individualization, there are numerous reasons for resisting change.

A primary problem for teachers facing laboratories for the first time is management. How does one get equipment out to 10 working groups, instruct them on certain fine points of operation, monitor equipment use, and get the equipment back washed and ready for the next class, all in 50 minutes? How does one have students carry out individual projects of various degrees of complexity while keeping a flow of continuing class work? Generally, how does one anticipate problems and cope when students are expected to be self-directed?

Many teachers find laboratory preparations frightening and inconvenient. One must know how to dilute concentrated sulfuric acid without a thermal explosion, know that hot paraffin is not to be poured down a sink drain, and know that agar slants are to be sterilized before washing.

Handling students in small- and large-group discussion in which the flow of questions and answers is driven by open-ended laboratory experience, grading laboratory books that are the reflection of what actually was observed in the laboratory and not what a text says should occur, using performance tests instead of paper-and-pencil tests—all can create apprehension and frustrations for the teacher.

For teachers not endowed with fix-it, scrounger, or entrepreneurial genes, inadequate facilities and equipment become insurmountable hurdles, and inadequacy is common.

Most teachers, when starting up a new laboratory-based curriculum, are frightened of questions they cannot answer. It takes long study to gain a sense of security about new content, and still greater stress is placed on teachers when new content is quantitative, rather than descriptive.

Training and Its Problems

The literature on educational change is clear. The kinds of problems listed above can be overcome only in intensive training. This point is sufficiently well accepted that federal agencies funding development and dissemination of new curriculum, such as the National Science Foundation (NSF) and the National Diffusion Network (NDN), now require developers to make training available to prospective users. But even here there are problems.

There are serious difficulties with present NSF- and NDN-like training practices. Training *made available* is not training *required*. A new package with optional training does not carry the message to administrators and teachers that successful implementation *depends* on the understanding of content, the inquiry style, the mechanics of the laboratory and field experience, and other subtleties.

Teacher training is normally done over vacations, when schools are not in operation. As a nonstandard activity of most schools, training is inconvenient and expensive. Teachers must give up well-deserved vacation time. School boards and administration must find money to coordinate the training, pay personnel and get them to the training site, maintain the site, etc. Training goes against the basic instincts of the cost-conscious administrator and school board.

Even when training is entered into, it is often insufficient in particularity and intensity. First, the training for curriculum change may be generic, dealing with a range of new content, teaching strategies, and laboratory techniques, but never dealing with the details of the curriculum that a teacher will face in the next term. Second, program-specific training may be delivered by a person who has no experience in teaching the curriculum to be used and who is therefore unable to speak to the problems that will have to be confronted in the classroom.

Seldom understood and even less often supported is the need for in-service coaching and mentoring after a training workshop is completed. In the best in-service training, only some of the potential problems of a curriculum can be touched on, and it will take several years for a teacher to master sufficiently the delivery of a given new curriculum to feel truly at home with it. In the early days and years of adjusting to a new curriculum, the help, counsel, mediation, and problem-solving of a creative mentor often are the difference between the curriculum's succeeding and failing.

Political Forces Frustrating Change

In their public zeal to establish curricular requirements to reflect some vision of curricular adequacy and currency and to ensure that those requirements are met, state legislatures nationwide have been erecting structures that work against their own intent to effect change. Such structures include rigid subject-matter and grade-level syllabi, mandated requirements, and testing. Stipulation of what must be taught reduces the potential for change. This is particularly true when requirements are tied in with paper-and-pencil testing.

Administrative Frustration with Change

Administrators are held accountable by the public for prudent budgetary management and the quality of teaching in their schools. For them, new curricula that make different or greater demands on resources, change the content and skill preparation of students, or change the definition of adequacy and excellence for the performance evaluation of teachers can be a nightmare.

Changes in textbooks are fairly easy to justify with school boards. Board members know that textbooks wear out and must be replaced. Changing to a laboratory-intensive program from a text-dominated program or from one laboratory program to another presents a very different scope of financial outlay and a very different kind of justification. Resulting budgetary requests can be defended only on the grounds that students are gaining a better understanding of science, and such a defense is very often hard to make, particularly when the state testing program does not reflect laboratory experience.

Every time there is a substantive change in the content and skill expectations of a course, external adjustments must be made. When the direction of the change is toward an inquiry laboratory program, many students who have done well in the past using texts will evidence early frustration and antagonism.

The professional skills demanded of a laboratory inquiry teacher are very different from those demanded of a textbook lecture teacher. For the administrator who is looking for the first time at a laboratory in which students are expected to assemble and operate experiments with minimal teacher input, the scene may appear chaotic. Pattern in activity is hard to detect, conversation among partners will range from cooperative and restrained to argumentative and unmodulated, and rates of getting down to task will vary markedly. Rating teachers' performance in such an environment is difficult. Most difficult is the situation in which a teacher rated as excellent as a lecturer is now struggling with a new program.

THE CRDG MODEL

Despite all the mechanisms for change that have been developed, countervailing forces have tended to reduce educational progress to a mime walk with great apparent movement and little forward progress. The same forces that have stymied us in the past exist today. In science, the problem is exacerbated, because teacher shortages are again bringing into the classroom teachers who are only marginally prepared for their assignment. What, then, can be done? Some possible answers come from the experience of the Curriculum Research and Development Group (CRDG) of the University of Hawaii, which over the last 22 years has been developing a series of techniques that deal with the problems outlined here.

CRDG is a semiautonomous unit within the College of Education. It has been mandated to serve the curricular and other educational needs of primary and secondary schools of Hawaii and the Pacific Basin. Its charge is to engage in curriculum research, development, dissemination, and evaluation.

Resources

Resources of the group include the following:

- The University Laboratory School, which acts as the primary test site for new programs. The Laboratory School has some 360 students,

K-12, who are selected from the four public-school districts on the island of Oahu to represent the socioeconomic, ethnic, and intellectual mix of students in the state.

- A permanent faculty of some 60 persons, augmented by a temporary staff of about equal size hired to complete particular projects.
- An annual budget of approximately $2,000,000, with additional funds generated out of grants and contracts from public and private schools in the state and Pacific Basin.
- Access to the services of disciplinary faculty members of the university who act as consultants, overload staff, or joint appointees.

Research

CRDG has a continuing research function that has three foci:

- *Screening.* New programs in selected curricular areas and grade levels are screened as they appear on the national and international scene and, when found promising, are visited, trial-tested, or otherwise studied. This activity has several outcomes. It may provide information for later curriculum design, provide a basis for advising schools on use, or provide the contacts for a CRDG role in program dissemination.
- *Exploratory research.* New curricular and administrative ideas generated by the staff are constantly being explored. For example, at this time, exploratory research is being done on ways to make all course offerings accessible to heterogeneously grouped classes; to combine the study of physics and physiology; to achieve problem-solving mastery in chemistry with computer monitoring and generation of problems; to define more clearly the learning behavior of students in their acquisition of algebra concepts; to service the special learning problems of the Pacific Island students making the transition to Hawaii's schools; to achieve more objective grading of student school performance; and to conceptualize, organize, and provide leadership for a program of prevention for students at risk and others.
- *Program effectiveness.* There is continuing research accompanying curricula already in dissemination and those in development. This includes formative evaluation during the early stages of laboratory school and pilot-testing and more classical summative evaluation during field-testing.

The University Setting Advantage

Though CRDG is product-oriented, there is expectation that considerable time will be spent in doing research. Research results can be weighed and validated, colleagues can be consulted, whole systems challenged and reconstructed, and ideas explored, often long before there is a general expression of need. Most important, there is an opportunity to explore the frontiers of ideas and a recognition that, of the many ideas explored, only a few will result in products.

In contrast, efforts to improve curriculum, such as those driven by federal funding, work out of expectations of success within the defined period of the grant. This means that efforts must be very circumscribed and circumspect. Substantive change requires a free atmosphere to think, tinker, and test—the atmosphere of the university—and few school systems can provide such an opportunity.

Situated outside the schools of the state, CRDG has been able to take on a range of topics with broad innovative content that could not be undertaken within the normal structure of the public or private schools. For example, in science alone, CRDG has conceptualized, designed, developed, tested, and disseminated the 3-year middle-school or intermediate-school integrated science program Foundational Approaches in Science Teaching (FAST); the K-12 Hawaii Nutrition Education (HNE) program; the nation's only 1-year high-school laboratory-based oceanography curriculum, the Hawaii Marine Science Studies (HMSS) project; and many others.

Development and Trial Procedures of CRDG

Targets for development may be identified by CRDG staff or by public or private schools in Hawaii or by schools or educational organizations in the Pacific Basin. When CRDG initiates a project out of the results of its own research, it does so only after consultation with the Hawaii Department of Education, which is its principal client. Once a new project is started, the following general steps are followed:

- The project is endowed with a staff, usually under the leadership of a senior faculty member.
- A steering committee is recruited and charged.
- Design is begun, with the steering committee as a sounding board and the laboratory school as a site for preliminary trial of ideas.
- Development proceeds to a full-scale laboratory school version that is tested, revised, and retested until deemed ready for pilot-testing.
- Piloting takes place in a selected group of schools with feedback going to revision of the materials.
- Field-testing and dissemination with in-service coaching and mentoring follow, along with regular testing and revision.

The Dash Model of Development and Dissemination

Of the dozen programs now in design and development, the K-6 Developmental Approaches in Science and Health (DASH) program has a structure that potentially offers solutions to some of the problems described above. Young children, who best understand concrete, immediate things, need science materials that reflect their home, school, and community environments. There is also a need to satisfy special state and local curricular requirements. Such needs cannot be accommodated in curriculum fabricated from afar. Therefore, a central core of materials is being developed in Hawaii and then pilot-tested, modified, and complemented by the staffs of eight collaborating mainland university schools. If successful, such a model may hold promise for other efforts to adjust prefabricated curricula to local needs. In addition, once developed, these materials will be disseminated and serviced by these same local university schools.

The FAST Model of Dissemination

CRDG and its science section in particular have had exceptional success in getting programs instituted and retained in schools. A specific example will give insights into our general approach. The FAST project was first pilot-tested in Hawaii in 1970. From the beginning, Hawaii teachers using the program were required to undergo an intensive training workshop, originally 6 weeks and eventually refined to 2 weeks. Teachers are supported by a field coordinator, who provides a variety of followup mentor services and sustains a collegiate community among FAST teachers. After 18 years, the training workshop still takes teachers through all activities of the program while instructing them in classroom management, as well as the program's philosophy and pedagogy. Where originally instructors were developers, they are now practicing teachers selected for their exemplary teaching of FAST and their capacity to communicate with their fellow teachers. Recent estimates indicate that, of the more than 500 Hawaii teachers trained in FAST who are still teaching middle-school science, some 90% are still using the materials.

National Dissemination

In 1984, CRDG received a grant from the National Diffusion Network (NDN) that enabled it to explore national dissemination of FAST. A marketing system was set up through the university's nonprofit research corporation, and a field representative was recruited to act as sales agent. All parties agreed to the following operational rules:

- No teacher is to be provided FAST materials until he or she has been trained in a registered FAST workshop. Once trained, a teacher is given a certification number. Orders for materials must be accompanied by the certification number or an agreement for training. The certification number is the property of the teacher.
- Trainers must qualify as University of Hawaii instructors, and university credit is given to workshop participants when desired.

- Schools are contracted with to provide an in-service followup contact person and continuing contact with the project.
- Teacher-training costs for individuals are borne out of costs of the FAST materials starter set.
- All training of trainers and the assignment of trainers are under the supervision of CRDG.

After 4 years, well over 1,000 FAST teachers have been trained in the continental United States, and they are teaching some 100,000 students this year. The teacher retention rate is about 90%, paralleling the Hawaii experience. The 4-year period has been a time of research for CRDG, and, although the model works in all its service aspects, there is yet question as to whether CRDG curricula less well known than FAST can succeed with the same mechanism of dissemination.

Cost

It is interesting to look at the developmental cost of FAST to get some notion of what price the state has to pay for tailor-made curriculum service. Over its 22 years of development and dissemination, the project has cost the state approximately $800,000. Over that same period, 200,000 of Hawaii's students have used the program at a cost of $4 per student. On the basis of an average expenditure of $2,400 per year per student, FAST has cost the state about 0.16% of the yearly outlay per student served.

CRDG Service

In the normal course of school service elsewhere, curricular consultation, conceptualizing and theorizing, exploratory research, design and development, and dissemination and mentor-coaching are done by different entities, if at all. In Hawaii's case, CRDG provides all these services, thus eliminating most of the confusion that comes when there is a multiplicity of service agencies. The net efficiency of this holistic system is much greater than that of the normal fragmented approach.

CONCLUSIONS AND RECOMMENDATIONS

As one draws conclusions about science-curriculum change in America, six points should be accommodated. First, a great strength of our educational system is its diversity and responsiveness to local need. Second, there is a plethora of institutions and agencies involved in curricular change, and often their methods and motivations for change run at cross purposes and may conflict with the needs of teachers, who are the ultimate institutors of change. Third, to accommodate all the changes required by legislatures and boards of education, much curriculum has become an unconnected patchwork of pieces without integrating logic. Fourth, although there is emphasis on curricular change, schools generally have no group to turn to that has continuously monitored the process of change and has intentionally sought answers to the question, "Where should we go from here?" Fifth, curricular packages have limitations as to how large an educational region they can serve without modifications. Sixth, at the level of classroom implementation, today's teachers are the most poorly prepared to teach science since World War II and have a desperate need for long-term in-service mentoring and coaching, if they are to provide quality science education. In sum, one is forced to the conclusion that a huge task of localizing curriculum and training teachers faces us, if we are to resolve the crisis of the eighties and provide the next great leap in biology education.

Out of the CRDG model comes a possible way of building on the strength of diversity and providing consultation, research, planning, localized curricular materials, and teacher training and coaching. To preserve diversity, it is suggested that a group of state or federally supported university-based educational institutes be created to devise and support new curricula. To achieve needed service levels, these institutions should be given six tasks:

- To monitor and research international and national, as well as local, science in some defined service area.
- To reflect continuously on and explore new curricular directions.
- To provide consultative services to legislatures, boards, and schools.
- To design and modify materials as needed within the service area.
- To provide in-service training and followup coaching and mentoring for teachers in their service area.
- To make the materials produced available to other service areas.

It has been the CRDG experience that teachers, administrators, and the various parties to the politics of education need external institutions commissioned to work on the spectrum of problems they separately and jointly face. These institutions need to have the independence to create and explore promising new ideas and to think holistically. Any new biology initiative will be well served by such a structure.

Index

A

B

E

F

J

K

L

M

N

O

P

Q

R

S

U

V

W

Y